Annals of the
American Conference
of Governmental
Industrial Hygienists

Volume 10

TD
883.
2

Evaluating Office Environmental Problems

AMERICAN CONFERENCE OF GOVERNMENTAL INDUSTRIAL HYGIENISTS

CINCINNATI, OHIO
1984

The Annals series will present the latest state-of-the-art information on research and practical applications of science in the field of occupational health. Each issue will be the proceedings of an important symposium, conference, or meeting sponsored by the American Conference of Governmental Industrial Hygienists or other leading professional organizations in, or allied with, the occupational health field. Contents will deal with subjects of current interest. The Annals series will become a valued addition to the international scientific literature.

Volume 1: Dosimetry for Chemical and Physical Agents (1981)
Volume 2: Agricultural Respiratory Hazards (1982)
Volume 3: Protection of the Sensitive Individual (1982)
Volume 4: ACGIH Transactions — 1982 (1983)
Volume 5: Industrial Hygiene — The Future (1983)
Volume 6: Computerized Occupational Health Record Systems (1983)
Volume 7: Some Pioneers of Industrial Hygiene (1984)
Volume 8: ACGIH Transactions — 1983 (1984)
Volume 9: Threshold Limit Values — Discussion and Thirty-Five Year Index with Recommendations (1984)

Copyright © 1984 by the American Conference of Governmental Industrial Hygienists, Inc. All rights reserved. Individual readers are permitted to make fair use of the Annals and the materials in them for teaching or research, under the provisions of the United States Copyright Act of 1976. Isolated articles covered by the Copyright may be photocopied in limited quantity for nonprofit classroom or library reserve use by instructors and educational institutions. Quoting from the Annals is permissible, provided that acknowledgment of the source is included. Republication of any material in the Annals is prohibited, except with permission of ACGIH. Inquiries should be addressed to the Executive Secretary, American Conference of Governmental Industrial Hygienists, 6500 Glenway Avenue, Bldg. D-5, Cincinnati, Ohio 45211.

Library of Congress Catalog Card No.: 84-70956

Printed in the United States of America

ISBN: 0-936712-52-X

CODEN: ACGHD2 10 1-136 (1984)

Annals of the American Conference of Governmental Industrial Hygienists

VOLUME 10

October 1984

Evaluating Office Environmental Problems†

Conference Chair:
John J. McFeters

CONTENTS

Probelm scope

3   Indoor air quality — the NIOSH experience
JAMES MELIUS, M.D., KENNETH WALLINGFORD, M.S., RICHARD KEENLYSIDE, M.D., and JAMES CARPENTER
National Institute for Occupational Safety and Health

9   Evaluating office environmental problems: consultant's overview of problem
EDWARD J. SWOSZOWSKI, Jr.
Consultant

15   Sources of air contaminants in the office environment
G. LYNN HOLT, CIH
Tennessee Valley Authority

21   Environmental studies in moldy office buildings: biological agents, sources and preventive measures
PHILIP R. MOREY, MICHAEL J. HODGSON, WILLIAM G. SORENSON, GREGORY J. KULLMAN, WALLACE W. RHODES, and GOVINDA S. VISVESVARA
Division of Respiratory Disease Studies, NIOSH; Rhodes Consultants, Inc.; Center for Infectious Diseases, CDC; Department of Medicine, University of Pittsburgh

37   Epidemiologic investigation of office environmental problems
DEAN B. BAKER, M.D., MPH
School of Public Health, University of California — Los Angeles

45   Application of health standards and guidelines
VERNON L. CARTER, Jr., D.V.M.
College of Veterinary Medicine, The Ohio State University; Chair, Chemical Substances TLV Committee

Office building ventilation: design

49   Ventilation concepts for office buildings
PRESTON E. McNALL, Jr. and ANDREW K. PERSILY
Center for Building Technology, National Bureau of Standards

59   Ventilation for acceptable indoor air quality: ASHRAE Standard 62-1981
JOHN E. JANSSEN
Honeywell, Inc.

---

† This series of papers is the result of a symposium entitled *Evaluating Office Environmental Problems* held by the American Conference of Governmental Industrial Hygienists on March 5-7, 1984.

Office building ventilation: assessment

69   Tracer gas measurements of ventilation in occupied office spaces
DAVID T. GRIMSRUD
Building Ventilation and Indoor Air Quality Program, Applied Science Division, Lawrence Berkeley Laboratory

77   Measurement of HVAC system performance
JAMES E. WOODS, Jr., Ph.D., P.E.
Physical Sciences Center, Honeywell, Inc.

93   Comfort and discomfort in office environmental problems
JOHN A. CARLTON-FOSS, Ph.D., S.M.
President, Human-Technical Systems, Inc.

Special ventilation problems

113   Commercial/institutional indoor air quality study by the Bonneville Power Administration
ROBERT ROTHMAN
Bonneville Power Administration, U.S. Department of Energy

115   Cross contamination and entrainment
RICHARD W. GORMAN, CIH
National Institute for Occupational Safety and Health

121   Case presentations: problems caused by moisture in occupied spaces of office buildings
PHILIP R. MOREY
National Institute for Occupational Safety and Health

129   A typically frustrating building investigation
STANELY A. SALISBURY
National Institute for Occupational Safety and Health, Region IV

Symposium Program Committee

GEORGE A. CARSON, Ph.D., CIH
National Institute for Occupational Safety and Health, Region VII

RICHARD A. KEENLYSIDE, M.D.
Rhode Island Department of Public Health

JOHN J. McFETERS
Tennessee Valley Authority

KENNETH M. WALLINGFORD
National Institute for Occupational Safety and Health

Recognition should be given to the members of the Symposium Program Committee for their hard work and insight which produced the program and identified the speakers for this symposium. Appreciation is also given to their employers for the support provided during the months of preparation.

The American Conference of Governmental Industrial Hygienists (ACGIH) believes it has a responsibility to provide an open forum for discussion of scientific questions. The positions taken by the participants in the reported portions of this text are their own and not those of ACGIH. ACGIH has no intent to influence legislation by providing such forums.

ANNALS OF THE
AMERICAN CONFERENCE
OF GOVERNMENTAL
INDUSTRIAL HYGIENISTS

VOLUME 10

EVALUATING OFFICE ENVIRONMENTAL PROBLEMS

PROBLEM SCOPE

Indoor air quality — the NIOSH experience

JAMES MELIUS, M.D., KENNETH WALLINGFORD, M.S., RICHARD KEENLYSIDE, M.D., and JAMES CARPENTER
National Institute for Occupational Safety and Health

Introduction

Health complaints associated with poor indoor air quality have been reported with increasing frequency among office workers over the past 10 years. A number of factors have contributed to this trend, including changes in building design to increase energy efficiency, introduction of new building materials, and increased awareness among workers of potentially toxic exposures in their work and home environments. While some exposures in the indoor environment may have serious implications for the health of the occupants (e.g., radon daughters, asbestos, formaldehyde, pathogenic microorganisms, allergens), the health risks of most indoor air exposures are poorly understood. Nevertheless, office workers who are concerned about a possible problem in their work environment often demand a thorough investigation of the environment and of its potential for health risks. For the investigator, this is commonly a challenging but frustrating experience because of the lack of guidelines and evaluation criteria for the "non-industrial" setting.

The health complaints reported by the occupants of the typical "problem building" are usually diverse and non-specific, and rarely point to an obvious cause. Also, the nature of the problem may be obscured by reports of a variety of serious, but unrelated, medical conditions associated with the office environment.

Appropriate environmental sampling is often quite difficult because of the presence of low levels of many ubiquitous substances. Specific contaminants from sources in the office can be readily measured and quantitated (e.g., ozone from copying machines), but more often very low levels of a variety of chemicals are detected. Such sampling results may in one sense be reassuring, but they have limited usefulness when linked with non-specific symptoms. These investigations are often complicated by extensive media coverage, multiple earlier investigations by other groups, and deteriorating labor relations. The "Federal expert" often comes into the picture when, despite much basic work having been done, the situation has reached a "crisis."

This paper will briefly present and review the indoor air quality investigations conducted by the National Institute for Occupational Safety and Health (NIOSH) since the start of the Health Hazard Evaluation Program. These investigations are being presented not as "THE WAY" to evaluate such problems, but, rather, to review our experience and share our insights as we have evolved our approach to these investigations.

Indoor air quality health hazard evaluations

Through December 1983, NIOSH has completed 203 Health Hazard Evaluations involving indoor air quality (IAQ) in a variety of settings (Table I). (This does not include our investigations of asbestos-related problems in office buildings.) Prior to 1978, only six IAQ evaluations were performed; however, since then, the number has increased dramatically. It appears that in the last 2 years, the number of these completed investigations has leveled off, but this change may reflect our handling of many IAQ inquiries by providing written materials and phone consultation, and by the increased capability of state and local health de-

TABLE I
Completed NIOSH Indoor Air
Quality Investigations by Year
(through December 1983)*

Year	Number Completed	%
Pre-1978	6	3.0
1978	9	4.4
1979	12	5.9
1980	28	13.8
1981	80	39.4
1982	44	21.7
1983	24	11.8
Total:	203	

* Does not include 81 currently active projects.

partments and other groups to handle these evaluations without our assistance.

Most of our evaluations have involved government and private offices (over 75% [see Table II]), educational institutions (14.8%), and health care facilities (9.3%). Given our mandate to evaluate occupational health problems, it is not surprising that NIOSH has not investigated very many residential IAQ problems. Thus, we do not have much experience in evaluating problems which are principally encountered in residential buildings such as exposures to radon daughters or to combustion products.

In reviewing the reports on these evaluations, we have attempted to classify our findings by the type of problem found (Table III). It is of note that nearly half of these investigations have attributed the IAQ problems to inadequate ventilation. Some form of environmental contamination was thought to be the source of the problem in approximately 30%. The source of this contamination was thought to be from inside the building in 17.7% of the investigations, outside the building in 10.3%, and from the building structure in 3.4%. Problems such as hypersensitivity pneumonitis, cigarette smoking, humidity, etc., have accounted for approximately 10% of our evaluations. Finally, in another 10%, the etiology of the IAQ problem has remained unexplained.

In reviewing these results, several factors should be considered. First, over time, NIOSH has not used a standard protocol for conducting these evaluations. Our methods and criteria have changed as we became more familiar with the problem and developed new approaches. Also, some of these investigations were conducted several years ago, leaving only scanty data and a brief report for current review. In the early studies, many of the "unknown" problems may have actually been due

TABLE II
Completed NIOSH Indoor Quality Investigations by Building Type
(through December 1983)

Type	Number	%
Government and business officer	154	75.9
Schools and colleges	30	14.8
Health care facilities	19	9.3
Total:	203	

TABLE III
Completed NIOSH Indoor Air Quality Investigations by Type of Problem
(through December 1983)

Problem	Number	Total
Contamination (inside)	36	17.7
Contamination (outside)	21	10.3
Contamination (building fabric)	7	3.4
Inadequate ventilation	98	48.3
Hypersensitivity pneumonitis	6	3.0
Cigarette smoking	4	2.0
Humidity	9	4.4
Noise/illumination	2	1.0
Scabies	1	0.5
Unknown	19	9.4
Total:	203	

to inadequate ventilation, but the reports did not provide enough information to determine this. Thus, there may be considerable misclassification in this list.

This listing is also not necessarily representative of the general distribution of indoor air quality problems in offices. Often, NIOSH is requested to conduct an evaluation only after initial attempts to identify the problem have failed, or complaints have persisted after initial corrections have been made. Large public-sector agencies use NIOSH as a resource, but managers of smaller offices or office buildings may be unaware of our program. Therefore, these facilities may be underrepresented on our list. Despite these shortcomings, the list does provide an overview of the types of indoor air quality problems encountered in office environments. A brief review of the major types of IAQ problems follows.

Contamination from inside the office environment

This classification (approximately 18%) refers to exposure to a chemical or other toxic agent generated within the office space. Usually the symptoms experienced by the office worker are directly linked to the exposure, but, if the exposure is disseminated through the building's ventilation system, localization of the source may be difficult. Examples from Health Hazard Evaluations include exposures to methyl alcohol from spirit duplicators,[1] exposures to methacrylate from copiers,[2] and exposures to sulfur dioxide from a heating system.[3] Occasionally, a very specific health effect can aid in the investigation, such as the occurrence of dermatitis from amines used in humidification systems.[4,5] The use of pest control agents, such as chlordane, may also cause persistent problems in office environments.[6,7]

These problems can usually be identified by inspecting the affected office area (e.g., copying machines) and by questioning about other uses of chemicals in the building (e.g., pesticides, humidification agents). Environmental and medical testing appropriate for that chemical can then be conducted.

Contamination from outside the building

This classification (approximately 10%) refers to exposure to a chemical or other toxic substance originating from a source outside the building. Common examples include motor vehicle exhaust either from a parking garage or loading dock entering the building through the intake from the ventilation system.[8,9] Many of these exposures are "presumed" explanations that cannot be documented at the time of the investigation. Although the dispersal of substances through the ventilation system may seem obvious, the complexity of the system may require the use of more sophisticated techniques such as a tracer gas to document the problem.[8] Other outside sources include nearby construction activity which may generate sufficient exhaust fumes, dust, or other contaminants to cause complaints in nearby offices.[10]

A few years ago, NIOSH evaluated a dramatic case of outside contamination of an office building when we discovered a 6-million-gallon underground gasoline spill while investigating irritative symptoms among the office occupants.[11] While this may be viewed as unusual, there is concern about the increasing number of underground gasoline tank leaks.

Contamination from the building fabric

Contamination from the building fabric (approximately 3% of our evaluations) refers to problems from the material used to construct the building. While not included in our list of evaluations, asbestos is obviously a major source of concern about indoor air quality. Other insulating materials are a common source of these problems because of the release of substances such as formaldehyde.[12] Dermatitis due to fibrous glass has also been a common problem, usually after the fibrous glass insulation has been disturbed during some construction activity.[13]

Hypersensitivity pneumonitis

This group (approximately 3% of our evaluations) refers to problems due to a hypersensitivity reaction to microorganisms in the building environment. Although not a common cause of office problems, the potential medical severity of this condition and the difficulties in controlling this problem make it an important cause of office problems. NIOSH's evaluations regarding this problem will be discussed later in this volume.

Inadequate ventilation

By far the largest classification, this group makes up approximately one-half of our completed evaluations. Our determination of inadequate ventilation is commonly made after considering a number of factors, including the absence of other sources of contamination and the presence of only very low levels of contaminants in our environmental sampling results. These, linked to the widespread occurrence of non-specific symptoms such as headaches, eye irritation, and upper respiratory irritation, suggest that an evaluation of the ventilation system may be warranted.

The evaluation of ventilation systems will be discussed later. Our methods range from obtaining specifications on the building ventilation system to detailed air flow measurements. Both approaches can present difficulties. Information on the specifications of the ventilation system is often not readily available from the building operators. Buildings with renovated ventilation system are often very difficult to evaluate because of the piecemeal approach used in such renovation. The ASHRAE guidelines for ventilation are usually used for comparison.[14] While one can question the basis for these guidelines, we have found them useful in evaluating IAQ problems and for recommending corrections where there is a problem because of inadequate ventilation.

The pathogenesis of complaints or symptoms caused by inadequate ventilation is not clear, but certain extreme situations have provided us with some insight into the relationship between such complaints and inadequate ventilation. In 1982, we conducted an evaluation at a government office building in Idaho with widespread complaints among the employees. Despite environmental surveys showing no significant comtaminant levels, the employees were moved to another building. Our investigators found that the air intake for the building had been covered with plastic one year earlier, to protect the air handling system from airborne debris from Mount St. Helens, and that this cover had never been removed. Removal restored intake of adequate outdoor air to the building and allowed reoccupancy without significant problems.[15] However, in this situation, no environmental measurement indicated that there was a problem. In general, we have not found any environmental measurement to be useful as an indicator of poor ventilation. However, it should be noted that other investigators have found carbon dioxide levels useful for such evaluations.[16,17] Low levels of multiple contaminants are often present in these situations and are currently the best explanation for the occurrence of symptoms. The pattern of contaminants probably varies from building to building, but we do not yet have adequate measurement techniques or adequate knowledge to easily recognize this problem through environmental sampling.

Current evaluation methods

Our current approach to evaluating office environment requests usually begins with a walk-through evaluation by an industrial hygienist. Prior to this, we try to obtain background information on the history of the building design or construction and try to ensure that the building engineer will be present during the investigation. During the initial visit, we obtain a history of complaints among office occupants by interviewing as many as is feasible. This is helpful not only for identifying the type of medical complaints, but also for obtaining a chronology of the problem and ascertaining the time pattern of symptoms (afternoon more than morning, etc.).

Potential sources of contamination are identified during the initial walk-through evaluation. Some will be obvious (e.g., copying machines), while others may be identified only after careful questioning (e.g., pesticide spraying). We also usually inspect the ventilation system for the particular office area and attempt to understand its connection to the system for the entire building. Information on the control of the ventilation system (outdoor air intake relative to temperature, etc.) is also obtained. Some environmental sampling may be conducted if a source of contamination is found or suspected. Some general air monitoring (e.g., organics) may also be conducted, but this is usually more helpful to reassure the occupants that the toxic substances of concern to them are not present in any degree than it is for identification of a problem.

The problem may be resolved during this initial visit, but, in some instances, more extensive environmental sampling, a medical study, a ventilation assessment, or some combination of these may be required. These may be necessary either to better identify the source and extent of the problem or to alleviate the concerns of the affected employees. Once our investigations are complete, our findings and recommendations are communicated to the involved parties.

Future activities in indoor air quality

More research into office ventilation and its effect on background levels of contaminants is necessary to provide better guidelines for evaluating and controlling indoor air quality problems. Because of the nature of these exposures, a variety of governmental agencies and private groups are involved in this research effort. Recently, the Environmental Protection Agency, the Consumer Product Safety Commission, the Department of Energy, the Department of Health and Human Services, the Tennessee Valley Authority, and several other Federal agencies formed a coordinating committee on indoor air quality research. This committee will help to coordinate Federal government indoor air quality research. Already, an inventory of IAQ-related research in the Federal government has been prepared. These agencies are also coordinating the planning of a possible large national survey of indoor air pollution and related health problems.

While much of this effort and related research may seem remote to investigating specific problems in office buildings, they may provide the basis for better guidelines for evaluating indoor air quality and for necessary corrective steps. At the same time, improved methods of assessment are needed to evaluate specific types of problems.

NIOSH is currently developing better methods for assessing indoor air quality (e.g., microorganism levels, ventilation parameters, etc.). Other groups are working on methods for other types of assessments. Meanwhile, we will continue to evaluate indoor air quality problems and, hopefully, continue to improve our efforts in these assessments.

References

1. Centers for Disease Control: Methyl Alcohol Toxicity in Teacher Aides Using Spirit Duplicators — Washington. *Morbidity & Mortality Weekly Report* 29:437-438 (1980).
2. National Institute for Occupational Safety and Health: *Congressman Cavanaugh's Office.* Health Hazard Evaluation Report No. HETA 80-067-754. NIOSH, Cincinnati, OH (1980).
3. National Institute for Occupational Safety and Health: *Wappingers Falls School.* Health Hazard Evaluation Report No. HETA 83-172-1409. NIOSH, Cincinnati, OH (1984).
4. National Institute for Occupational Safety and Health: *Boehringer-Ingelheim, Ltd.* Health Hazard Evaluation Report No. HETA 81-247-958. NIOSH, Cincinnati, OH (1981).
5. National Institute for Occupational Safety and Health: *Cornell University.* Health Hazard Evaluation Report No. HETA 83-020-1351. NIOSH, Cincinnati, OH (1983).
6. National Institute for Occupational Safety and Health: *Georgetown University.* Health Hazard Evaluation Report No. HETA 83-444. NIOSH, Cincinnati, OH (under study).
7. National Institute for Occupational Safety and Health: *Planned Parenthood.* Health Hazard Evaluation Report No. HETA 83-258. NIOSH, Cincinnati, OH (under study).
8. National Institute for Occupational Safety and Health: *Department of Justice.* Health Hazard Evaluation Report No. HETA 80-024-887. NIOSH, Cincinnati, OH (1981).
9. National Institute for Occupational Safety and Health: *Cincinnati Technical College.* Health Hazard Evaluation Report No. HETA 82-269-1341. NIOSH, Cincinnati, OH (1983).
10. National Institute for Occupational Safety and Health: *McCalls Publishing Company.* Health Hazard Evaluation Report No. HETA 81-097-1021. NIOSH, Cincinnati, OH (1981).
11. Centers for Disease Control: Employee Illness from Underground Gas and Oil Contamination — Idaho. *Morbidity & Mortality Weekly Report* 31:451-453 (1982).
12. National Institute for Occupational Safety and Health: *Tri-Valley Federal Credit Union.* Health Hazard Evaluation Report No. HETA 81-108-883. NIOSH, Cincinnati, OH (1981).
13. National Institute for Occupational Safety and Health: *Ellis Hospital.* Health Hazard Evaluation Report No. TA 80-080. NIOSH, Cincinnati, OH (1980).
14. American Society of Heating, Refrigerating and Air-Conditioning Engineers, Inc.: *ASHRAE Standard 62-1981, Ventilation for Acceptable Indoor Air Quality.* ASHRAE, Atlanta, GA (1981).
15. National Institute for Occupational Safety and Health: *U.S. Forest Service.* Health Hazard Evaluation Report No. HETA 81-150-994. NIOSH, Cincinnati, OH (1981).
16. Rajhans, G.S.: Indoor Air Quality and CO_2 Levels. *Occupational Health in Ontario*, pp. 160-167. P.L. Pelmear, Ed. Ontario Ministry of Labour (1983).
17. Hicks, J.B.: Tight Building Syndrome: When Work Makes You Sick. *Occup. Safety & Health*, pp. 51-57 (1984).

Evaluating office environmental problems: consultant's overview of problem

EDWARD J. SWOSZOWSKI, Jr.*
Consultant

While the many factors and intricacies of this new "science" of indoor air quality continue to evolve and demand definition, it is already obvious that more than the traditional approaches and methodologies will be required to adequately deal with indoor environmental situations. Influences beyond medical, scientific, and engineering considerations weigh heavily on the development of programs designed to identify and resolve problems in buildings.

Government regulations and standards define procedures in dealing with only a small number of the situations encountered in buildings. Even when requirements and guidelines do exist, there is often doubt as to their adequacy or relevance to the non-occupational conditions found in buildings. This doubt is fueled by a fear of potential litigation and liability. Medical and scientific fact are often no match for the emotions, fears, and rights of facility users as defined by the media and the courts. A consultant operating in this volatile environment must do so without the protection that "sovereign entity" affords to government agencies. Such vulnerability often leads to the realization that any attempt to deal with indoor environmental problems must include disciplines and concepts beyond the scope of industrial hygiene and safety.

Only recently has an understanding of the many independent factors which define a building's environment been combined with a knowledge of the materials and practices which contribute to indoor environmental problems.[1,2] In some instances problems can be quickly defined and resolved. Other attempts to deal with indoor situations only reveal shortcomings which prevent a timely and comprehensive solution.

To illustrate the potential and real problems faced by a client and consultant in dealing with indoor environmental situations, I would like to address the example of asbestos-containing materials in buildings. I do so for two reasons:

1. It is the office situation with which I am most familiar.
2. It mirrors well the potential problems and liabilities surrounding an attempt to safely and efficiently deal with a potentially hazardous material in buildings.

At first note it would appear that concern over the presence of asbestos-containing materials in buildings would not include some of the questions and lack of supporting information which surround other materials in the indoor environment.

- Asbestos is a recognized human carcinogen.
- No safe level of exposure has been established. Indeed, a recent U.S. Consumer Product Safety Commission funded study was very clear in its recommendation that "the Commission should regard asbestos at all levels of exposure as a potential human carcinogen."[3]
- Occupational Safety and Health Administration (OSHA) regulations dealing with asbestos-related activity already exist. OSHA proposed changes and NIOSH recommendations address possible shortcomings in the current regulation.
- A U.S. Environmental Protection Agency (EPA) Asbestos in Schools program offers recommendations and guidance for handling asbestos-related situations in schools and other building types.

Yet, despite the existance of this pool of information, regulations, and recommendations, attempts to deal with the potential problem of asbestos-containing materials in schools and other facilities have met with only limited success. Some would not be so kind. "A dismal failure..."

---

* P.O. Box 561, South Worwalk, Connecticut 06854; 203/866-1141.

"Caused more harm than good . . ." Such comments are often heard among those who have taken the time to investigate the results of hazard identification and asbestos abatement programs.

The reasons for this failure not only relate to asbestos, but have ramifications for all individuals who must deal with a potentially harmful situation in buildings. I would like to present seven observations for your consideration.

1. Existing federal and state regulations rarely address the non-production or non-occupational exposure situations found in office buildings. Sampling technology and methodology defined in such regulations are often unable to accurately define potential exposure situations in office buildings. Such factors as brief but intense exposure situations, sampling methodology inappropriate for building environments or activities, and the possibility that levels of contamination are below the detection limit of the testing method hinder assessment efforts.

In situations where a regulatory agency does recognize the limitations of existing technology and methodology, as does the EPA in strongly recommending that air monitoring not be used for assessing the need for corrective action, no specific information, guidelines, or alternatives are offered to accurately determine potential hazard.

At this point any recommendation a consultant may make to a client, save for complete elimination of the problem, involves both risk and liability for both. The courts have already ruled that compliance with OSHA regulations does not absolve an employer from liability in cases of worker health and safety.[4] Other decisions have questioned the ability of "state-of-the-art" programs to adequately protect workers. Is an "all or nothing" solution the only alternative open to a facility operator who is faced with a potentially hazardous condition?

Ignoring or expressing reluctance to deal with a potentially hazardous situation in a building does not constitute responsible action. Yet my experience with asbestos has demonstrated that even sincere and timely attempts to deal with a problem by completely removing asbestos-containing material from a building are not without significant problems.

Until a comprehensive regulation, based upon current medical and scientific research on the potential for exposure to asbestos in non-industrial situations, is created and implemented, facility operators, consultants, abatement contractors, and the general public will suffer from the threat of risk, liability, and possible exposure to a recognized human carcinogen.

2. No strong, effective, visible, and uniform regulatory entity exists to administer the planning, execution, and enforcement of existing regulations and standards. Interpretation and enforcement of regulations vary from jurisdiction to jurisdiction, promoting inconsistencies in information and recommendations on work-related practices and procedures. This very real lack of consistency only increases the confusion of facility operators and other involved parties whose major concern is only "to do the right thing."

Because of this uncertainty in interpretating and enforcing regulations, the range of building operator response in dealing with asbestos-related situations has been staggering. Two recent examples from the North East illustrate the point.

In one New England state no standards exist within any of its state departments that define safe and efficient asbestos-related abatement. The current federal OSHA occupational standard (2.0 f/cm^3, 8-hr TWA) is used consistently to define risk in schools.[5-7] Indeed, the director of the state OSHA has stated that even in a situation where bulk asbestos material had fallen to the floor and retarded children were eating it, the situation "didn't present any hazard."[8]

In a large New England city, a portion of the ceiling in a retail clothing establishment was covered with an asbestos-containing material. The material was in good condition. Air monitoring conducted by qualified professionals defined extremely low (0.02 f/cm^3) levels of airborne fiber. Yet the local health department, without benefit of accurate and demonstratable methods of defining real or probable risk, required the material to be removed. The abatement project cost the building owner and tenant in excess of several hundred thousand dollars, with no guarantee that the end result provided the public with a building any safer than existed originally. The local health department showed no interest in adequately inspecting or testing the facility after abatement to determine the quality of the abatement effort. Attempts to have the department define what standard will be used in the future to assess asbestos-in-building situations have been ignored or brushed off with "politically safe" responses.

Who is correct? And is any effort at abatement necessary if the end result is a less safe condition than preceeded the corrective effort?

A major recommendation of the EPA's asbestos guidance document is that of having building operators themselves create detailed specifications for asbestos-related identification and abatement programs. In this way they might avoid the shortcomings in the current occupational asbestos regulations. Facility operators are also encouraged to provide the effort necessary to ensure specification compliance.

The shortcomings of such a recommendation are not difficult to imagine. The quality of specifications varies considerably, from those which are comprehensive and accurate, to others which reflect individual preferences rather than medical, scientific, or engineering fact . . . or even common sense. As a result:

- Specifications have been ignored to the point that significantly more problems were created that solved.
- Abatement contractors have filed significant claims against clients based upon losses or hardships resulting from inaccurate or incomplete specifications.
- Contractors and clients alike have started legal proceedings against architects and consultants for inappropriate or incomplete specifications.

It is difficult to imagine that any building operator or consultant has at their disposal research and enforcement resources the equal of those available to Federal and state government agencies. Yet the burden of creating comprehensive and accurate specifications, which not only must deal safely and efficiently with a recognized human carcinogen but also prevent risk and liability, is given to such individuals.

The existance of regulations dealing with non-occupational asbestos-related activity fulfills only a portion of the current assessment and resolution requirements. Without strong, effective, uniform, and accurate interpretation and enforcement of such a regulation performance shortfalls will continue to exist, and workers and facility users will continue to be exposed to asbestos.

3. While current federal regulations do define the proper methods for the disposal of asbestos-containing waste, for clients and consultants alike the task is complicated by differing site conditions, requirements, and levels of regulation enforcement. State and local regulations only add to the confusion.

4. Attempts for the Federal government to provide for a dialogue between regulatory agencies and those involved in performing asbestos-related activity have had only limited success. Federal agencies have not been able to transfer the findings of medical, scienfitic, and technological research to building operators, consultants, and contractors in a meaningful and effective manner. The lack of governmental real world experience, and the failure to utilize such experience from non-governmental sources is primarily to blame.

In the case of friable asbestos materials, the alarm was sounded by the government on the basis of medical and scientific research demonstrating potential exposure of building users. Little thought was given to the actual problems which might be encountered by building operators and contractors dealing with such material during the course of maintenance, renovation, or abatement activity. This oversight has had wide ranging ramifications for building operators and contractors alike. Exposure of workers and building occupants to asbestos has occured during attempts at corrective action prompted by concerns raised by the EPA's guidance documents. Without proper information and guidance in the techniques and standards for safe asbestos-related activity, problems have often been compounded rather than solved. Almost without exception contractors entering the field of asbestos-related activity have had to undergo a "trial and error" learning period to develop their capability. Such periods of experimentation have often included less than safe trial situations.

Government sponsored studies, while advancing science and the state-of-the-art understanding of asbestos-related subjects, offer little "hands-on" information for building operators and contractors. Operators and contractors have had to learn for themselves, often at great risk to workers and facility users, that EPA recommended procedures do not work equally as well on types of asbestos. If the goal of these guidance documents is to develop a "read it once and go out and do a safe and complete job" document for real world use, then the government agencies have failed.

Without clearly defined and on-going channels of communication between regulatory agencies and those groups directly involved with building environmental situations, the effectiveness of any program dealing with hazardous substances in buildlings is questionable. Government researchers are not going to directly deal with building situations themselves. That will be left to real world operators, consultants, and contractors. Until we can create the proper tools for these individuals, and provide them with accurate information concerning the task at hand, errors, accidents, and exposure situations will continue to occur.

Recognition is needed of the value of participation of such groups as facility operators, architects, consultants, contractors, facility workers and staff, and the public in expanding the data/experience bases, and in helping identify potentially hazardous situations in buildings.

5. Inappropriate methodology and the lack of understanding of office activities, maintenance, renovation, and abatement activity have often led to inappropriate or incorrect air monitoring data.

A cornerstone of most asbestos programs and regulations is the monitoring of levels of airborne asbestos fiber. Since the regulations were created to deal with industrial situations, the collection methodology reflects the often stationary, repetitive nature of that kind of work. Such collection techniques have little relevance to office or corrective action activity, where potential levels of airborne fiber are often intense and short-term.

Without proper, timely, and efficient air monitoring by qualified technicians/microscopists, air monitoring data cannot be considered accurate or useful in contributing to a safe building environment in situations involving asbestos.

6. Perhaps the most discouraging to consultants and contractors involved in asbestos-related activity is the lack of owner awareness concerning the requirements of such activity.

Facility operators are often not aware of the level of sophistication of techniques, protective equipment, and procedures necessary for safe and effective asbestos-related maintenance, renovation, or abatement activity. They are often stunned at the cost of these requirements as compared to normal building services.

Competent contractors bidding on the same comprehensive asbestos-related specification, and knowing that specification requirements will be enforced, will be competitive in price while maintaining quality standards. In situations where specifications are lax, or specification enforcement is weak, contractors often cut corners, and therefore price, by compromising on protective safeguards, equipment, or proper technique. A low price is attractive to any facility operator. Sadly, that is sometimes the only criteria used to select a contractor for asbestos abatement activity.

Building operators and those involved in paying for corrective action would be well advised to follow the recommendation of the EPA concerning asbestos abatement.

"*Successful abatement, not cost miniminization, is the goal. A premium may well be justified to help assure success.*"[9]

One has only to consult with legal counsel about asbestos-related liability and toxic torts to begin to appreciate the legal quagmire which can be avoided by timely, responsible, and comprehensive action concerning asbestos in buildings. The same holds true with other potentially hazardous building situations.

Sadly, many government entities and private facility operators are locked into a low-bid requirement either because of regulation or by choice. While such an approach works well in normal service acquisition situations, it has little value in those cases where a potentially hazardous building problem must be defined and resolved by competent and informed vendors. A general low bid requirement can be used only after all potential bidders have been screened for competency and ability using a well defined contractor pre-qualification standard.[10]

    7. Perhaps the least discussed problem affecting consultants, contractors, and building operators involved in asbestos-related activity is the question of insurance and bonding.

Insurance carriers bear an enormous portion of the potential liability associated with asbestos-containing materials. Claims by individuals who have incurred asbestos-related diseases along with those of the asbestos producers and distributors have greatly increased insurance industry liability. One economist has estimated that insurance industry liability related to asbestos will reach $40 billion by 1990.[11] As a result, insurance carriers are understandably gunshy at the prospect of dealing with non-industrial asbestos-related activity.

Building operators have been known to describe asbestos-related activity to their insurance carrier as simply "renovation activity" to avoid higher premiums or the risk of obtaining no insurance coverage at all. In other cases a contractor's bonding and insurance carrier had no idea that the "renovation" activity for which they were providing coverage included asbestos-related activity. On more than one occasion an insurance carrier has not found out the truth until the small weekend "renovation" turned into a several hundred thousand dollar, building-wide decontamination project.

Several large insurance carriers have dropped all contractors engaged in asbestos-related activity from bonding and insurance coverage. Others are reluctant to accept consultants or architects engaged in such activity, citing a lack of industry standards and liability data. While that relieves insurance industry concerns, one must ask the question, "If no coverage is available, then who will attempt the necessary asbestos-related remedial work?" Or is it the intent of the insurance industry, the government, and the courts to place the whole issue on a back burner while standards are created, liability data collected, and the insurance industry lobbies for government reinsurance for their potential asbestos-related liability.

Without the direct and on-going particiaption of the insurance industry in the formation and application of standards and requirements for non-industrial asbestos-related activity building operators, consultants, and contractors will be reluctant to engage in such activity to the detriment of those workers and facility users subjected to asbestos exposure situations. Current insurance industry response to such reservations is uncertain and hesitant. Few carriers desire to create precidents which might endanger current legal proceedings or increase risk on existing coverage. The carriers must realize that cooperating with government and private researchers to create standards and oversee performance will, over a period of time, reduce their potential liability by not only controling abatement activity but reducing the general building population currently exposed to asbestos.

It is uncertain how long it will take the federal government to create a new asbestos standard which will adequately protect workers and the general public in building situations. Undoubtedly the asbestos industry, insurance carriers, and their legal representatives will take an active part in the debate over any proposed regulation or standard which might threaten their economic viability. So to will the litigation attorneys, labor organizations, interest groups (such as the PTA, tenant organizations, building owners and operators) and contractors and consultants who will actually deal with the corrective process.

Somewhere into the proceedings the medical and scientific communities will admit that while they can accurately describe the nature and mechanics of asbestos-related disease, little is truly known about asbestos-related activity in buildings and the potential for exposure and disease resulting from such programs. Such shortcomings will necessitate research into the problem. Meanwhile, more and more time is lost; more building operators will enter into the emotional and complicated problem of what to do about asbestos-containing materials in their buildings. The "tensions" provided by facility users, interest groups, and litigation attorneys will demand immediate action. And sadly the result will probably involve the exposure of even more people to asbestos.

Asbestos will be the first in a long line of substances which will contribute to the evolution of toxic tort precedents. How we go about creating the technology and standards to deal with asbestos-containing materials in a safe and efficient manner will not only create precedents in medicine and science, but will have a profound influence on how the general public perceives the actions of manufacturers, insurance carriers, building operators, contractors, consultants, labor organizations, government agencies, and the courts in responsibly protecting their health and safety. We should not allow the issue to degenerate into a "regulation by litigation" scenario. Any regulation, standard, or action pertaining to asbestos should be based upon the best possible medical, scientific, and engineering fact.

David Hull, president of the Philosophy of Science Association was quoted in an article entitled "Is Science Stymied by Today's Complexity?" as stating:

> " 'In general the simple problems have all been solved. . . . perhaps the most difficult of all are the delicate questions of human health. The challenge is not only to define 'health' in the first place but also to look for effects throughout a lifetime, possibly decades after exposure.' "[12]

Any action to define potential risk in a non-occupational asbestos-related situation, and control exposures in such cases must be accurate, responsible, and timely in we are to keep the public's confidence. The role of the consultant in situations involving building conditions and asbestos containing materials is far from defined. In this expertless field, where fact and demonstrable risk can be displaced by emotion, fear, and legal precedent the role best played by a consultant is as part of a multidiscipline team. Only by including legal and insurance understanding to medical, scientific, and engineering expertise can any such building problem be adequately addressed.

We must act responsibly in protecting the health and safety of building occupants and staff. But we must be prepared not only to go beyond traditional medical and scientific criteria, but also be ready to explain and defend the decisions and actions made to achieve that goal.

References

1. Meyer, B.: *Indoor Air Quality.* Addison-Wesley Publishing Company, Inc., Reading, MA (1983).
2. Spengler, J.D. and K. Sexton: Indoor Air Pollution: A Public Health Perspective. *Science 221(4605)*:9-17 (July 1983).
3. Chronic Hazard Advisory Panel on Asbestos: Report to the U.S. Consumer Product Safety Commission. U.S. Consumer Product Safety Commission, Directorate for Health Sciences, Washington, DC 20207 (July 1983).
4. Barley vs Bell Asbestos. Eastern District Court of Pennsylvania (January 1981).
5. Despite Asbestos Precautions Parents Worry. New Haven, CT *Register.* (October 13, 1983).
6. Asbestos Level High in 6 Schools. *Ibid.* (June 21, 1983).
7. No Asbestos Danger. New Haven, CT *Journal-Courier.* (February 2, 1984).
8. School Clean-up Urged. Hartford, CT *Courant*/United Press International. (February 2, 1984).
9. U.S. Environmental Protection Agency: *Guidance for Controlling Friable Asbestos-Containing Materials in Buildings.* EPA 560/5-83-002. Office of Pesticides and Toxic Substances, Washington, DC 20460 (March 1983).
10. Swoszowski E.J.: *Considerations in Selecting an Asbestos Abatement Contractor* (in press) (1983).
11. National Council on Compensation Insurance: *The Asbestos Crisis — Yesterday, Today, and Tomorrow* (1983).

Sources of air contaminants in the office environment

G. LYNN HOLT, CIH
Tennessee Valley Authority

Introduction

Twenty-five percent of America's employees work in an office environment. An increasing number of these office employees seem to be experiencing work-related health problems for which traditional industrial hygiene techniques have often been found to be inadequate. Traditional industrial hygiene techniques, which rely on recognition, evaluation, and control, break down when there are literally hundreds of potential sources of contaminants any of which or combination of which may be causing an employee's symptoms. The fact that these symptoms may appear at exposure levels hundreds or even thousands of times below appropriate occupational standards also strains normal industrial hygiene procedures.

Although this paper will be limited to only the discussion of the identification of office contaminants and sources, this is but one of the many confusing problems confronting industrial hygienists called on to solve office environmental problems. Some of these problems are shown in Table I.

The problems caused by indoor contaminants have become more prevalent within the last 10 years. One primary cause has been the need to reduce the amount of energy required for heating and cooling office buildings. The method most often used to reduce this energy consumption has been to "tighten" the building to prevent the gain or loss of air which requires energy to be conditioned. To accomplish this, many building owners have also closed or constricted the inlet to the makeup air duct, caulked cracks, and insulated surfaces as much as possible to reduce energy consumption and cost. Tightly closed buildings prevent the removal and dilution by outside air of the pollutants generated by an increasing variety and quantity of synthetic materials introduced into the indoor environment. Indoor pollutants build up faster than they can be diluted resulting in occupant discomfort and complaints of stuffiness, eye irritation, fatigue, drowsiness, nausea, and stale air.

TABLE I
Causes of Frustration for the Industrial Hygienist Evaluating Office Environmental Problems

1. Contaminants and their sources are not always obvious.
2. Only a few employees may complain of a problem.
3. Traditional industrial hygiene measurements usually fail to identify cause.
4. Occupational health standards are not appropriate for the office environment.
5. Problems may often be cyclical or episodic in nature.
6. It may be difficult to distinguish between building-related symptoms and those of "normal" colds or respiratory infections.
7. Situation may have already become emotionally charged before industrial hygiene assistance is requested.
8. Management reactions may range from being overly sensitive to skepticism.

TABLE II
Sources of Contaminants Outside Building

Sulfur dioxide	Bacteria
Particulate	Ozone
Pollen	Lead
Carbon dioxide	*Carbon Monoxide
Fungi	

Sources

These contaminants can come from sources located outside the building, the building materials, building equipment, building maintenance and cleaning materials, or the building inhabitants and the products they use. Personal experiences in locating sources of contaminants have involved ammonia from microfiche equipment, mold from old files, acetic acid from engineering drawing copying machines, and probably the most difficult source to control — tobacco smoke. Contaminants and their sources for five general categories are presented in Tables II through VI. Only the more significant contaminants (identified in the tables by an asterisk and listed in Table VII) will be discussed below.

Carbon monoxide (CO), the most toxicologically significant of the products of combustion of fuels,

TABLE III
Building Material Sources of Contamination

Contaminant	Sources
*Formaldehyde	Particle board
	Urea-formaldehyde foam insulation
	Pressed wood
	Plywood resins
	Hardwood panelings
	Carpeting
	Upholstery
*Asbestos	Draperies
	Filters
	Stove mats
	Floor tile
	Spackling compound
	Furnaces (older)
	Roofing
	Gaskets
	Insulation
	Acoustical Material
Organic vapors	Carpet adhesives
	Wool finishes
*Radon decay products	Concrete
	Brick
	Stone
	Soil
Man-made mineral mibers	Fiberglass insulation
	Mineral wool insulation

has been reported to have reached elevated levels when released into the indoor environment from such sources as unvented gas appliances and heaters, malfunctioning heating systems, kerosene heaters, and underground or connected garages. Carbon monoxide is a lethal gas which is responsible for many fatalities each year. Carbon monoxide released from faulty heating systems, attached or underground garages, or tobacco combustion may easily exceed allowable occupational limits for short periods of time. CO ties up the hemoglobin of the blood preventing normal oxygen uptake and thus deprives the body of oxygen. This reduced oxygen supply is especially damaging to the brain and heart.

Formaldehyde is an irritating and possibly carcinogenic vapor emitted from many products found in the office environment. Sources of formaldehyde include urea-formaldehyde foam insulation (UFFI), particle board, pressed wood, plywood, carpeting, upholstery, and tobacco smoke. Up to 20% of the population may be unusually susceptible to eye irritation caused by formaldehyde vapors at concentrations of 0.25-0.5 parts per million (ppm).

Asbestos fibers are a common office contaminant because of their widespread use in asbestos-containing products including but not limited to gaskets, filters, spackling compounds, and acoustical and thermal insulating materials. This is especially true of buildings constructed between 1940 and 1970. Asbestos exposure has been linked to increases in lung cancers, pleural and peritoneal mesotheliomas, and cancers of the gastrointestinal tract. In addition, possible synergism with other contaminants (tobacco smoke) suggests exposure to asbestos should be as low as possible. The release of asbestos fibers is extremely difficult to monitor due to its episodic nature. Maintenance work, renovation, negligence, or vandalism may result in high airborne concentrations of asbestos fibers for a relatively short duration.

Radon decay products may be one of the most insidious of the indoor contaminants. Radon is a radioactive gas emitted naturally by building ma-

TABLE IV
Building Equipment Sources of Contamination

Contaminant	Sources
Ammonia	Microfilm machines
	Engineering drawing reproduction machines
Ozone	Electrical equipment
	Electrostatic air cleaners
*Carbon monoxide	Combustion sources including: gas ranges, dryers, water heaters, kerosene heaters, fireplaces and wood stoves, garages
Carbon dioxide	
Surlfur dioxide	
*Nitrogen dioxide	
Hydrogen cyanide	
Particulates	
Benzo(a)pyrene	
Amines	Humidification equipment
Carbon	Photocopying machines
Methyl alcohol	
Trinitrofluorene	
Trinitrofluorenone	
Methyl alcohol	Spirit duplication machine
Methacrylates	Signature machine
Dust	Video display terminals

TABLE V
Maintenance Material Sources of Contamination

Contaminant	Sources
*Formaldehyde	Paper towels
	Disinfectants
	Carpet shampoos
Sodium carbonate	Commercial cleaning products
Sodium perborate	
Sodium phosphates	
Ammonium compounds	
Borax	
Pine oil	
Trichloroethylene	
Naphtha	
Kerosene	
Petroleum solvents	
Alkyl benzene sulfonate	
Alkyl aryl sodium sulfonate	
Sodium dodecyl sulfate	Carpet shampo
Organics	
	Varnishes
Phthalates	Paints
Methylene chloride	Polishes
Benzene	Cleaners
Cumene	etc.
Benzaldehyde	
Mesitylene	
Linonene	
2-methyl-naphthalene	
Diethyl phthalate	
Tetradecane	
1,1,1-Trichloroethane	
1,1,2,2-Perchloroethylene	
Cyclohexane	
n-Heptane	
n-Octane	
n-Nonane	
n-Decane	
n-Dodecane	
n-Tridecane	
n-Pentadecane	
n-Hexadecane	
5-Penyldecane	
Toluene	
Ethylbenzene	
o,m,p-Xylene	
Propylbenzene	
1-Ethyl-2-methylbenzene	
1-Ethyl-4-methylbenzene	
C'-Alkyl benzenes	
Cyclohexanol	
2-Hexanone	
3-Heptanone	
Butylcyclohexane	
Naphthalene	
n-Butylacetate	
Limonene	

TABLE V (Continued)
Maintenance Material Sources of Contamination

Contaminant	Sources
Carvone	
1-Methylnaphthalene	
Ethylsilane	
Menthone	

terials such as concrete block and brick or soil which contain radium 226. The radon gas decays resulting in airborne elements which are also radioactive. These radon decay products tend to stick to airborne particles which then may be inhaled. Lack of adequate ventilation in a building may allow these products to accumulate to unnecessarily high levels. An increase of 1,000-20,000 lung cancer deaths may be a result of indoor exposure to radon decay products. The buildup of radon decay products is especially troublesome when air change rates per hour are less than 0.5 or about 5 cubic feet of fresh air per minute per person.

Tobacco smoke seems to be getting more and more attention by the news media and by many lawmaking bodies. Smoking restrictions (particularly in the office environment) are becoming more common. Almost 33% of the adult population smokes an average of 32 cigarettes each day. Most of these cigarettes are probably smoked on the job near fellow workers. Tobacco smoke has been proven to be a human carcinogen and respiratory toxicant. Tobacco smoke contributes to office air contaminant concentrations of particulates, nicotine, CO, NO_2, acrolein, polycyclic aromatic hydrocarbons, and an estimated 3,000 other substances.

Nonsmokers as well as smokers exposed to tobacco smoke experience adverse symptoms such as conjunctival irritation, nasal discomfort, cough, sore throat, and sneezing. These effects of tobacco smoke can usually be lessened if a fresh air supply rate of 20 cubic feet per minute per person is maintained in areas where smoking is allowed. However, maintaining this fresh air supply level may not be adequate to prevent the discomfort of some employees who are usually sensitive to tobacco smoke.

TABLE VI
Building Inhabitants as Sources of Contaminants

Contaminant	Sources
*Formaldehyde	Toothpaste
	Smoking
	Grocery bags
	Waxed paper
	Facial towels
	Shampoo
	Cosmetics
	Medicines
Acetone	Bioeffluents
Butyric acid	
Ethyl alcohol	
Methyl alcohol	
Ammonia	
Odors	
*Asbestos	Talcum powder
	Ironing board covers
	Hot mittens
	Hair dryers (old)
Acrolein	*Smoking
Nicotine	
CO	
+ 3000 others	
Dusts and vapors	Personal care products
Vinyl chloride	Aerosol spray, propellants, and solvents
Dusts	
Vapors	Cleaning products
	Hobbies
	Mothproofing
	Fire retardant
	Insecticides
*Radon decay products	Ground water
	Shower
Dusts	Fertilizers
Vapors	Adhesives
Polychlorinated biphenyls (PCBs)	Carbonless carbon paper

TABLE VII
Significant Contaminants

* Carbon monoxide	* Radon decay products
* Formaldehyde	* Smoking byproducts
* Asbestos	

have its own microenvironment. If the work area is found to have inadequate fresh air supply and this problem is corrected, this will probably remedy the contaminant problem without really identifying the source. All sources are important contaminants, but the lesser ones will drive you just as crazy. A brief protocol of investigation for evaluating office environmental problems is shown in Table VIII. Remember, employees will attribute illnesses to the workplace if they see excessive dust or poor housekeeping and if there is a peculiar odor, especially if they are susceptable to allergies.

TABLE VIII
Investigation Protocol

1. Characterize Symptoms and Complaints
   a. Distribution
   b. Location
   c. Time

2. Evaluate Ventilation System
   a. Quantity of fresh air per person
   b. Check air cleaning equipment
   c. Measure carbon dioxide and carbon monoxide levels.

3. Conduct a walk-through looking for obvious sources of contaminant.

Conclusion

Frequently when the source of the contaminant is not obvious it may be more reasonable to determine if the cause of the buildup of the contaminant was insufficient makeup air rather than attempting to identify every contaminant and determine if it is present in concentrations greater than permissible exposure limits. Unfortunately, sources and contaminants are probably going to be building specific since each building seems to

References

Alter, Dr. H. Ward: Statement for presentation before the Subcommittee on Energy Development and Application and the Subcommittee on Natural Resources, Agriculture Research and Environment of the Committee on Science and Technology of the U.S. House of Representatives (August 2, 1983).

Ayer, H.E. and D.W. Yeager: Irritants in Cigarette Smoke Plumes. *Am. J. Pub. Health 72(11)*:1283 (1982).

Bernstein, R.S., W.G. Sorenson, D. Garabrant et al: Exposures to Respirable, Airborne Penicillium from a Contaminated Ventilation System: Clinical, Environmental and Epidemiological Aspects. *Am. Ind. Hyg. Assoc. J. 44(3)*:161-169 (1983).

Budiansky, S.: Indoor Air Pollution. *Env. Sci. Tech. 14(9)*:1023-1027 (September 1980).

Budnitz, R.J., J.V. Berk, C.D. Hollowell et al: Human Disease from Radon Exposures: The Impact of Energy Conservation in Residential Buildings. *Energy and Buildings 2*:209-215 (1979).

Committee on Aldehydes, Board on Toxicology and Environmental Health Hazards, Assembly of Life Sciences, and National Research Council: *Formaldehyde and Other Aldehydes*. Academy Press, Washington, DC (1981).

Ehreth, Donald J.: Statement before the Subcommittee on Energy Development and Applications and the Subcommittee on Natural Resources, Agriculture Research and Environment Committee on Science and Technology, U.S. House of Representatives (August 3, 1983).

Florey, C du V, R.W. Melia, S. Chinn et al: The Relation between Respiratory Illness in Primary Schoolchildren and the Use of Gas for Cooking. III. Nitrogen dioxide, respiratory illness and lung infection. *Intl. J. Epidemiol. 8(4)*:347-353 (1979).

Freedman, T.: Warning — Staying at Home can be Dangerous to your Health. *Common Cause*, pp. 13-17 (1982).

Gesell, T.F., R.H. Johnson, Jr. and D.E. Bernhardt: *Assessment of Potential Radiological Population Health Effects of Radon in Liquefied Petroleum Gas*. U.S. Environmental Protection Agency, Office of Radiation Programs, Washington, D.C.

Hess, Charles T., Ph.D.: Testimony before Subcommittee on Energy Development and Applications and the Subcommittee on Natural Resources, Agricultural Resources and Environment.

Hicks, J.B: Tight Building Syndrome: When Work Makes You Sick. *Occupational Health and Safety* (January 1984).

Hileman, B.: Indoor Air Pollution. *Env. Sci. Technol. 17(10)*:469-472 (1983).

Hollowell, C.D., J.V. Berk, M.L. Boegel et al: *Building Energy Management*. Pergamon Press, Oxford, New York, Toronto, Sydney, Paris, Frankfurt (1980).

Honicky, R.E., C.A. Akpom and J.S. Osborne: Infant Respiratory Illness and Indoor Air Pollution from a Woodburning Stove. *Pediatrics 71(1)*:126-128 (1983).

Kane, D.N.: Bad Air for Our Children. *Environment 18(9)*:26-34 (1976).

Kirsch, L.S.: Behind Closed Doors — The Problem of Indoor Pollutants. *Environment 25(2)*:16-45 (1983).

Kreiss, K., M.G. G. Gonzalez, K.L. Conright and A.R. Scheere: Respiratory Irritation Due To Carpet Shampoo: Two Outbreaks. *Environment Intl. 8*:337-341 (1982).

Levin, L. and P.W. Purdom: A Review of the Health Effects of Energy Conserving Materials. *Am. J. Pub. Health 73(6)*:683-690 (1983).

Melius, James, M.D.: Testimony before Subcommittee on Energy Development and Applications and the Subcommittee on Natural Resources, Agriculture Research, and Environment, United States House of Representatives (August 3, 1983).

Moschandreas, Demetrios J.: Testimony on Indoor Air Quality Research Needs before the joint hearing of Energy and Development Applications and Environment Subcommittee (August 2, 1983).

Nero, Anthony V., Jr.: Statement before joint hearing of the Energy Development and Applications and Natural Resources, Agriculture Research and Environment Subcommittees of the House Committee on Science and Technology.

Pepys, J.: Chemical Dusts, Vapours, and Fumes causing Asthma. *Environment Intl. 8*:321-325 (1982).

Preuss, Peter W.: Statement before the Subcommittee on Energy Development and Applications, and the Subcommittee on Natural Resources, Agriculture Research and Environment, and the Committee on Science and Technology, of U.S. House of Representatives.

Rajhans, G.S.: *Indoor Air Quality and CO_2 Levels*.

Replace, J.L.: Indoor Air Pollution. *Environmental Intl. 8*:21-36 (1982).

Riley, R.L.: Indoor Airborne Infection. *Ibid. 8*:317-320 (1982).

Spengler, J.D., D.W. Dockery, W.A. Turner et al: Long-term Measurements of Respirable Sulfates and Particles Inside and Outside Homes. *Atmos. Env. 15*:23-30 (1981).

Spengler, J.D., C.P. Duffy, R. Letz et al: *Env. Sci. Tech. 17(3)*:164-168 (1983).

Spengler, J.D., B.G. Ferris, Jr., D.W. Dockery and F.E. Speizer: Sulfur Dioxide and Nitrogen Dioxide Levels Inside and Outside Homes and the Implications on Health Effects Research. *Env. Sci. Tech. 13(10)* (October 1979).

Spengler, J.D. and K. Sexton: Indoor Air Pollution: A Public Health Perspective. *Science 221(4605)* (July 1983).

Sterling, T.D. and E. Sterling: Carbon Monoxide Levels in Kitchens and Homes with Gas Cookers. *J. Air Poll. Cont. Assoc. 29(3)*:236-241 (March 1979).

Stolwij, J.A.J.: Indoor Air Quality Research: *Current Status and Future Needs*. Testimony before a joint hearing of the Subcommittees on Energy Development and Application and the Natural Resources and on Agriculture Research and Environment (August 3, 1983).

Thun, M.J., J.F. Lakat and R. Altman: Symptom Survey of Residents of Homes Indulated with Urea — Formaldehyde Foam. *Env. Res. 29*:320-334 (1982).

Traynor, G.W.: Pollutants Emissions from Portable Kerosene — Fired Space Heaters. *Env. Sci. Tech. 17*:369-371 (1983).

Tribble, Joseph J.: Statement before the Subcommittee on Energy Development and Applications and the Subcommittee on Natural Resources, Agriculture Research and Environment of the Committee on Science and Technology, House of Representatives (August 3, 1983).

Woods, James E.: Summary of Testimony presented at U.S. House of Representatives Committee on Science and Technology joint hearings of the Energy Development and Applicaiton Subcommittee and the Natural Resources, Agricultural Research and Environment Subcommittee.

Woods, J.E.: *Sick Building Syndrome — The Search for a Cure.*

Airtight Homes Pose Health Hazard in Pacific Northwest. *The News 60(7)*:1, 4 (October 17, 1983).

Better Assessment of Indoor Air Quality Needed to Identify Hazards — Official Says. *Occupational Safety and Health Reporter*, p. 296 (August 1983).

Building Mass, Ventilation Combine to Keep Upstate N.Y. Homes Comfortable. *Solar Energy Intelligence Report*, p. 382 (November 1983).

Fresh Air Without Frostbite. *Rodale's New Shelter*, pp. 58-63 (January 1984).

Indoor Air Pollution. *U.S. News and World Report*, p. 69 (October 1983).

OSHA: Fungus Flap in Fort Worth a Misunderstanding. *Occupational Health and Safety News*, pp. 10c-10d (November/December 1983).

Respiratory Illnesses Linked to Carpet Cleaning in Hospital. *Occupational Health and Safety Letter*, p. 8 (August 1983).

Ions in the Workplace. *Secretary 43(8)*:15-16 (October 1983).

Wood Stoves in 19% of the Homes. *Solar Energy Intelligence Report*, p. 390 (December 1983).

Environmental studies in moldy office buildings: biological agents, sources and preventive measures

PHILIP R. MOREY,[A] MICHAEL J. HODGSON,[A,D] WILLIAM G. SORENSON,[A] GREGORY J. KULLMAN,[A] WALLACE W. RHODES,[B] and GOVINDA S. VISVESVARA[C]

[A]Division of Respiratory Disease Studies, NIOSH; [B]Rhodes Consultants, Inc.; [C]Center for Infectious Diseases, CDC; [D]Department of Medicine, University of Pittsburgh

NIOSH has carried out health hazard evaluations in five large office buildings where hypersensitivity pneumonitis (HP) and other respiratory diseases have been alleged or reported. Environmental studies made both on the occupied space and in the heating, ventilation, and air conditioning (HVAC) system of each building are described. Several buildings were characterized by a history of repeated flooding and all contained mechanical systems with pools of stagnant water and microbial slimes. Preventive measures which may be effective in reducing building-associated microbial contamination and building-associated HP illnesses include the following: a) prevent moisture incursion into occupied space and HVAC system components; b) remove stagnant water and slimes from building mechanical systems; c) use steam as a moisture source in humidifiers; d) eliminate the use of water sprays as components of office building HVAC systems; e) keep relative humidity below 70%; f) use filters with a 50-70% rated efficiency; g) discard microbially damaged office furnishings; and h) initiate a fastidious maintenance program for HVAC system air handling units and fan coil units.

Introduction

Hypersensitivity pneumonitis (HP) and other respiratory diseases in office workers have been repeatedly described since 1970 (Table I).[1-8] Symptoms have included pulmonary manifestations such as chest tightness, coughing, and wheezing together with constitutional symptoms such as muscle aches, chills, fever, headache, and fatigue. Attack rates have varied from approximately 1%[4] to over 50%.[5] Disease has been attributed to thermophilic actinomycetes,[1-3] non-pathogenic amoebae,[7] fungi,[8] and *Flavobacterium* spp. or their endotoxins.[6] Despite extensive investigation the agents responsible for several large outbreaks were not determined.[4,5] In all outbreaks listed in Table I, the source of microbial contaminants responsible for the disease outbreak was thought to be a component of the building heating, ventilation, and air conditioning (HVAC) system. HVAC system ductwork,[1,3] humidifiers,[5-7] air washers (components of air handling units [AHUs] emitting a water spray),[2,4] and fan coil units (FCUs)[8] have been documented as sources of disease agents. Remedial and preventive measures used during HP outbreaks are poorly described, but have ranged from the cleaning of a HVAC system component[3,8] to total replacement both of the HVAC system[4] and all furnishing in the occupied space of the building (Table I).

Over the past 5 years, the Division of Respiratory Disease Studies of the National Institute for Occupational Safety and Health (NIOSH) through its Health Hazard Evaluation Program has been requested by other governmental agencies to evaluate apparent outbreaks of HP in office buildings. Since microbial agents are involved in the etiology of most HP-illnesses, environmental investigations in these buildings were primarily concerned with the identification of possible sources of biological agents. The suggestion of appropriate abatement procedures is an important component of each health hazard evaluation. Since October 1981, NIOSH has carried out environmental studies in five large, multistory office buildings wherein HP-like illnesses or other respiratory diseases were reported or alleged to occur. This paper describes environmental studies made both in the occupied space and in the HVAC system of each building. The possible sources of biological agents in each building are described and remedial and preventive measures which may be effective in controlling environmental contamination that has been associated with building-related HP are reported.

TABLE I
Sources of Microbial Contamination, Disease Agents, and Remedial Actions Associated with Past Outbreaks of HP or Allergenic Respiratory Disease

Author	No. Persons Ill/Total Exposed	Source of Building Contamination	Etiologic Agent	Remedial Action
Weiss & Soleymani[1]	Case report	Dust in HVAC system ductwork	Thermophilic actinomycetes	Tried unsuccessfully to clean occupied space; replaced HVAC system
Banaszak et al[2]	4/27	HVAC system water spray	Thermophilic actinomycetes	Unknown
Scully et al[3]	1/40	Dust in HVAC system ductwork	Thermophilic actinomycetes	Clean ductwork
Arnow et al[4]	48/4023	HVAC system water spray	Unknown	HVAC system replaced; all furnishings in occupied space replaced
Ganier et al[5]	26/50	Stagnant water in humidifier	Unknown	Tried unsuccessfully to decontaminate humidifier with fungicide; removed humidifier from HVAC system.
Rylander et al[6]	3/7	Aerosol from humidifier reservoir	*Flavobacterium* spp. or their endotoxins	Unknown
Edwards[7]	20/50	Microbial slime in reservoir	Nonpathogenic amoebae	Filter air entering humidifier; water in humidifier run to waste; replaced furnishings in occupied spaces
Bernstein et al[8]	2/14	Contaminated fan coil units	*Penicillium*	Clean fan coil units; replace filters

Sampling methods

Collection of airborne microorganisms was carried out using Andersen[A] viable samplers[9] including a modified Andersen sampler in which only the lowest sieve plate (stage 6) is utilized.[10] Total colony forming units (CFU)/m^3 were reported after correction of plate counts by the positive hole procedure.[9] Plates utilized in samplers for collection of fungi, bacteria, and amoebae contained rose bengal streptomycin agar (RBS) (100 µg streptomycin per ml), tryptic soy agar (TSA) (50 µg cycloheximide per ml), and nonnutrient agar coated with a suspension of *Escherichia coli,* respectively. Additional airborne microorganisms were collected in a two-piece cassette containing a sterilized cellulose ester filter (0.8 µm pore size) and backup pad as described elsewhere.[8] Andersen viable and filter cassette samplers were operated for a variable length of time at flow rates of 28.3 and 2.0 liters per minute (lpm), respectively. Bulk dust and water samples collected at building sites were analyzed for fungal and bacterial levels by standard serial dilution techniques using RBS and TSA media, respectively.

Respirable dusts were collected utilizing either a 1.27 cm (0.5 in) cyclone sampler (polyvinyl chloride 37 mm filter with nominal 0.8 µm pore

---

[A]Mention of company names or products does not constitute endorsement by the National Institute for Occupational Safety and Health.

size) operating at 9 lpm, or a 2.54 cm cyclone sampler (Bendix Model 240, polyvinyl chloride 47 mm filter, 0.8 µm nominal pore size) operating at a flow rate of 66 lpm.[11] Total suspended particulate was collected in occupied space utilizing a General Metal Works, 20 × 25 cm (8 × 10 in), high volume sampler (type AE glass fiber filter) operating at a flow rate of 2 m^3/minute.

The volume of outside air entering the mixing plenum of HVAC system AHUs in some buildings was estimated by multiplying the total area of louvered intake openings by the average air velocity through the free area of these openings. A rotating vane anemometer and/or an Alnor 6000 velometer was utilized to determine air velocity through open outside air intake louvers. The total volume of air flowing through each AHU was estimated at a position approximately eight cm downstream from the filter bank and was the product of the cross-sectional filter area and the average air velocity. The amount of conditioned air entering occupied space through ceiling diffusers was determined as the product of average air velocity (Datametrics hot wire air velocity meter) and the area of diffuser openings.

Short-term colorimetric indicator tubes including those for CO_2, CCl_4, O_3, NH_3, NO_2, CO, Cl_2, hydrocarbons, perchloroethylene, and methanol-ethanol were utilized to test for the presence of possible contaminant gases in occupied spaces. Other possible air contaminants were collected in a large charcoal tube, desorbed with ethanol and screened by gas chromatography. Environmental sampling for airborne fibrous glass was carried out by the fiber count method.[12]

Environmental studies in office buildings
Building A

At least one-third of the 350 employees in a building located in a southern city experienced recurring outbreaks of a febrile illness that led to a permanent evacuation of the facility. This building was constructed in the 1930s. Its HVAC system was installed in 1941 and contained two open water systems with nozzles for spraying water over a finless direct expansion evaporator coil for summer cooling. A mixture of return and outdoor air (up to 15%) passed through a filter bank and then through a water spray system in either of the two units. The spray water in each system was collected in a 12.7 m^3 (3000 gal) tank, chilled, and then reaerosolized. Conditioned air from each spray water-direct expansion system passed through baffle plates, into fans, and then was transported via ducts to occupied space throughout the building.

The HVAC system was turned off on September 18, 1981, as cool weather was expected the following week. Because the temperature in the building had reached 30°C (85°F) by 0800 hrs on September 21, the HVAC system was turned back on. During that day and evening an illness occurred in approximately one-third of the employees and consisted of headaches, muscle aches, fever, chills, nausea, wheezing, and chest tightness. In most affected individuals the symptoms had resolved by the next morning. Since there was a temporal relation between illness and turning the HVAC system on, the HVAC system was shut down and water spray systems were cleaned with steam and a quaternary ammonium compound. The HVAC system was then operated without effect upon building occupants until October 10 when it was shut down to repair one water spray system. The HVAC system, including water sprays, was turned on again on October 12 (Columbus Day, a holiday) and on October 13, a second mass-illness occurred. The HVAC system was turned off, and remained off until October 15. On that day the HVAC fan system was turned on, a third mass illness occurred, and building occupants were moved into other office facilities.

Review of building operations showed that outbreaks of mass illness could not be related to climatological conditions or to process variables such as new construction, renovation, or cleaning activities. Each instance of mass illness was however temporally related to the activation of the HVAC system.

The baffle plates of both air handling systems were coated with a slime growth. Similarly slimes were found on the surfaces of water spray sumps, on both pipe insulation and masonary wall surfaces located between water spray systems and fans, and on the floor in the vicinity of the fans. The microorganisms listed in Table II were isolated from these slimes. Slimes also contained unidentified flagellates, free living nematodes (e.g., *Rhabditis* spp.), and mites. Water spray sump waters were dominated by *Flavobacterium* spp.,

TABLE II
Microorganisms Found in Slimes in
HVAC System of Building A

Bacteria:	*Bacillus* spp.* (including 55°C thermophiles)
	Flavobacterium spp.*
	Pseudomonas spp.
	55°C Actinomycetes*
Fungi:	*Penicillium* spp.*
	Cladosporium spp.
	Cephalosporium spp.*
	Aspergillus spp.*
	Trichoderma spp.
	Mucor spp.
	Ostracoderma spp.
	Rhodotorula spp.
	Cryptococcus spp.
	Fusarium spp.
	Harposporium spp.
Protozoa:	*Acanthamoeba* spp.*
	Vorticella spp.
Nematodes:	*Rhabditis* spp.

* Organisms which have been implicated as etiologic agents in previous outbreaks of HP or humidifier fever.

amoebae (e.g., *Acanthamoeba* spp.), and ciliates (e.g., *Vorticella* spp.).

Levels of airborne microorganisms in occupied space of Building A were measured 10 days after the third outbreak of mass illness, at a time when the HVAC system had been turned off and water spray systems had been drained. Airborne fungal levels were approximately 800 CFU/m^3. At the same time respirable dust levels (1.27 cm cyclone) in occupied space were 46 μg/m^3, and the relative humidity was found to be 52%. However, eight hours after turning on the HVAC system including one water spray system, the relative humidity in occupied space rose to 78%.

The microbial contamination found in the HVAC system of Building A was considered to be responsible for the recurring outbreaks of febrile illness. A number of isolates from Building A including *Bacillus* spp.,[13] *Flavobacterium* spp.,[6] thermophilic actionmycetes,[1-3] *Cephalosporium* spp.,[14] *Aspergillus* spp.,[15] *Penicillium* spp.,[8,16] and *Acanthamoeba* spp.[7] have been implicated as agents in other outbreaks of HP and humidifier fever. If the source of the agent responsible for this disease outbreak was the aerosolization of microorganisms in or near the water sprays, the high relative humidity associated with this type of HVAC system operation would be conducive to the survival of viable particles at locations throughout the ductwork and in occupied space of this building. Consequently, it was recommended that all non-disposable building contents, including books, desks, carpets, drapes, and HVAC system ductwork and water spray-direct expansion system surfaces be thoroughly cleaned with a vacuum incorporating a high efficiency particulate air (HEPA) filter. It was further recommended that building contents which could not be adequately cleaned be discarded and replaced.

Building B

In this building, which is located in an eastern city, 12 of 41 office workers in a central zone on the seventh floor experienced HP-like symptoms consisting primarily of fever, chills, muscle aches, and chest tightness. Building B is an eight-story structure constructed in 1975 that is currently occupied by over 1300 employees. Most floors have eight office zones. One centrally located zone on the seventh floor has been the site of a series of floods, one of which occurred on January 27, 1982. Over the following months, several persons in this zone experienced a subacute febrile illness. Ill persons were more likely than non-ill persons to occupy desks within 5 m of water leaks.[17] The cafeteria kitchen on the eighth floor is directly above this zone and its water-drainage system runs through the common return air plenum above the suspended ceiling over offices occupied by personnel experiencing illnesses. The plumbing for the cafeteria dishwasher had no grease traps. Consequently, grease periodically clogged drain pipes, causing water to back up and eventually flood the underlying office zone.

Eighteen main AHUs and over 900 FCUs condition supply air in Building B. Main AHUs condition a mixture of outdoor and return air whereas FCUs condition only recirculated air. Conditioned air from each main AHU is transported to offices through ducts which terminate in slots around the periphery of ceiling light fixtures and in long slot type supply outlets at some building perimeter locations. Each AHU in Building B provides supply air to vertically superimposed zones on a number of different floors. Return air from occupied space passes into centrally located slots in ceiling lighting fixtures and enters the return air plenum. Return air from zones on each floor moves through this common plenum and then through shafts or risers, and is transported (by return fans) to main AHUs or is expelled from the Building during economizer operation. Air interchange between AHUs occurs because of mixing both in common return plenums and in building risers. Thus, once an air contaminant enters a common return plenum it could be easily distributed into other AHUs and then throughout the building.

Microorganisms were isolated from damaged ceiling tiles and carpets obtained from the seventh floor zone where illness had occurred. In addition, similar analyses were carried out from debris obtained from the outside surface of pipes in the common return air plenum above this zone plus a sample of water collected during a flood which occurred on April 1, 1982. All specimens examined for protozoa contained *Acanthamoeba polyphaga*. Other predominant microorganisms isolated were *Monosporium apiospermum*, *Rhodotorula* spp., *Aureobasidium* spp., *Colpoda* spp., and *Mastigamoeba* spp. Sampling for airborne fungi was carried out in this office on March 16 and again on May 7, 1982. On both occasions levels of fungi were low, being less than 100 CFU/m^3. Dust samples collected from main AHUs and FCUs throughout the building contained *Thermoactinomyces* spp. as predominant isolates. Attempts using serologic techniques to determine if *Acanthamoeba*, *Aureobasidium*, *Thermoactiomyces*, and other microorganisms isolated from Building B were disease agents proved inconclusive.[17] Additional microbiological studies revealed that up to 1×10^8 bacteria/ml were present in flood waters reaching the office zone where illness occurred. Stagnant water containing microbial slimes was present in the drain pans of some AHUs.

The average concentration of respirable dust collected in this office zone with the 2.54 cm cyclone was 25 μg/m^3. Total dust and CO_2 levels never exceeded 40 μg/m^3 (American Society of Heating, Refrigerating, and Air-Conditioning Engineers [ASHRAE] recommended limit is 260 μg/m^3 over a 24 hr period)[18] and 400 ppm (NIOSH recommended limit of 10,000 ppm),[19] respectively. All tests for contaminant gases by colorimetric indicator tubes were negative. Air samples collected by charcoal tubes contained toluene, xylene, and a series of mostly branched alkanes in the C_{10} to C_{12} region, but only in trace amounts, and were log orders below applicable occupational threshold limits.

Because disease was related to working within 5 m of water leaks[17] and because the affected office was contaminated with a variety of microorganisms, some of which are known to cause HP-lung disease,[17,20] it was postulated that the affected persons experienced respiratory exposure to an unidentified microbial agent associated with flooding from the overhead cafeteria.

Only minimal levels of airborne fungi were recovered by sampling on March 16 and May 7, 1982. However, on both instances aerobiological sampling was carried out at least one month after the office zone had been flooded, at a time when airborne levels of viable spores may have been dissipated or affected by such variables as temperature, humidity, clean-up activities in occupied spaces, and filtration by the HVAC system. In another office building where HP was shown to be caused by spores emanating from contaminated FCUs,[8] the time at which aerobiological sampling is carried out was shown to be critical in attempts to relate disease prevelence to airborne levels of microorganisms. Sampling conducted on the day that FCUs were operated in their heating mode showed that airborne fungal spore density was 50 to 80 fold above background levels; on the previous day when FCUs were quiescent, spores in occupied space were present only at background levels.[8] By analogy, aerobiological sampling in Building B might have been a more useful indicator of disease prevalence if it had been carried out during or within a few days after the floods of January 27 and April 1, 1982.

Clean-up measures recommended for the affected office zone in Building B included the following:

1. Structural changes in cafeteria plumbing should be made so that flooding in offices is prevented.

2. Discard damaged carpeting and ceiling tiles; clean the outside surfaces of pipes from which floods originated; clean all upholstered furniture, wall partitions, and office materials that need to be reused with a vacuum incorporating a HEPA filter.

3. Disinfect the floor with bleach and then refurnish and reoccupy the office.

After our investigation and during the clean-up of the affected office in the third week of May 1982, large amounts of dust were liberated when office partitions were handled. Illness reoccurred in previously ill individuals. Even though our analyses were unsuccessful in identifying the etiologic agent, it was probably present in office partitions in May. There are many potential reasons for the difficulty in identifying the specific disease agent, and these include:

1. The agent may be an organism other than the predominant ones that were isolated from the environmental samples.

2. The agent may not be viable and therefore was not cultured from bulk samples and was absent from the panel of antigens used in serologic studies. Disease may also have been caused by microbial toxins such as endotoxin.

3. The exact etiology of this disease may be demonstrable only by provocative challenge which was not attempted in this study.

Building C

Respiratory illnesses were studied among office workers in a nine-story building located in a southern city. One employee on the seventh floor of the building had symptoms and spirometry suggestive of HP. Two other workers in the same office appeared to have work-related respiratory symptoms. In August 1982 an industrial hygiene survey was carried out in this building to determine if there was an environmental basis for these complaints.

Conditioned air is supplied to office areas by a HVAC system that contains five main AHUs. Each AHU supplies air to an adjacent pair of floors. The unit supplying the seventh floor also provides air to the floor below. FCUs located next to outside walls on all floors provide supplemental heating and cooling in occupied spaces. Air from occupied space enters the common return plenum on each floor through slots in light fixtures, moves through a large riser along with return air from three or more floors, and is subsequently transported to AHUs by return fans. Therefore, in Building C, it is possible that a contaminant from one AHU might enter additional AHUs because of the mixing of return air streams.

The back of each AHU is located next to an outside building wall. A variable amount of outside air enters AHUs directly through louvered intakes and dampers in these walls. Outside air is combined with return air in a mixing plenum, the mixed air stream passes through a roll filter (no ASHRAE atmospheric dust spot efficiency rating),[21] and then successively through cooling coils, and an air supply fan. Cooled and dehumidified air (summer mode of operation) is transported through a system of ducts and into occupied space through slots found in the sides of some fluorescent light fixtures.

The cooling deck of the AHU serving the seventh floor lacked an adequate drainage pan necessary for the collection and removal of condensed water. As a result, condensed water pooled to a depth of 5 cm and stagnated on the deck of the plenum housing the cooling coils and the air supply fan. Several liters of microbial slime were found on the wetted surfaces of this AHU. Drain pans in FCUs

TABLE III
Fungi and Bacteria in Samples from the HVAC System and the Cooling Tower of Building C

Sample Description	Fungi/ml	Fungi/g	Bacteria/ml	Bacteria/g
Slime from AHU	2.0×10^5	—	7.4×10^7	—
Condensate water from AHU	1.3×10^2	—	1.2×10^5	—
Water from cooling tower	3.5×10^2	—	1.3×10^5	—
Moist sludge from condensate tray of fan coil unit	—	1.1×10^6	—	6.1×10^6

found in the offices occupied by complainants were also coated with a thick layer of slime.

Water and slime samples from the AHU and from one FCU were analyzed for viable fungal and bacterial contents. For comparison purposes, similar analyses were carried out on water collected from an external source of microorganisms, namely a cooling tower on the roof of Building C. Condensate water from the AHU and from the reservoir of the cooling tower contained equivalent numbers of microorganisms (Table III). However, slime from the AHU contained concentrations of bacteria and fungi several orders of magnitude greater than that characteristic of cooling tower water. Sludge from the condensate pan of the FCU was also more heavily contaminated with microorganisms. Cooling towers are well known microbial incubators.[22] Since contamination within AHUs and FCUs is equal to or greater than that found within cooling towers, and because microbially contaminated components in AHUs and FCUs are in close proximity to HVAC system air distribution fans, it is likely that microorganisms are aerosolized directly into the conditioned air stream supplied to occupied space.

The number of people working on the sixth and seventh floor of Building C was approximately 220. The total volume of conditioned air moving through the AHU serving these floors was 11,600 L/s (23,200 cubic feet/min (cfm)). Of the total air flow, 3075 L/s (6150 cfm) was outside air. Assuming equivalent distribution of conditioned air to all sixth and seventh floor offices, each occupant received about 14 L/s (28 cfm) of outdoor air, which exceeded the ASHRAE recommended value of 10 L/s (20 cfm) per occupant for buildings where cigarette smoking is permitted.[18] Additional air measurements were made in an office occupied by one of the complainants on the seventh floor. The office housed 6 employees (one case, 5 noncases) and was being provided with a total of 280 L/s (560 cfm) of conditioned air through slots around ceiling lighting fixtures. Assuming that 26% of the conditioned air being provided to the seventh floor is derived outdoors (11,600 L/s (23,200 cfm) total flow through AHU; 3075 L/s (6150 cfm) outdoor air; assume perfect mixing in AHU and in room), this office received 74 L/s (148 cfm) of outdoor air or approximately 12 L/s (24 cfm) outdoor air per occupant, again slightly above ASHRAE minimal recommendations.

Health related complaints offered by building occupants on the seventh floor are thus likely due to causes other than inadequate ventilation.

Tests for other air contaminants were negative. Levels of respirable dust collected (2.54 cm cyclone) in the return air stream were always less than or equal to 50 $\mu g/m^3$. Contaminant gases were not detected by short-term indicator tubes. A thick layer of eroded fibrous glass was found lining the inside structural surface of AHUs and the main air supply duct downstream from the fan. Microbial analysis of a dry sample of fibrous glass obtained from supply air ductwork indicated that approximately 200 viable fungi and bacteria were present in each gram of insulation. Although airborne fibrous glass particles in occupied space were present in trace levels (< 0.05 fibers/cc) far below the NIOSH recommended threshold of 3 fibers/cc,[12] it does appear likely that this deteriorating insulation harbors a large population of microorganisms which can be potentially aerosolized into occupied space.

The following were among the remedial recommendations made with regard to Building C:

1. Provide adequate drainage of condensate water under cooling coils. For drains originating in the vicinity of cooling coils, install deep sealed-water filled traps.[23] Trap depth should exceed the maximum suction pressure head created by the fan.

2. Clean and disinfect cooling coils and drain pans of AHUs and condensate pans of FCUs.

Building D

Following an outbreak of HP in the middle 1970s[4] this office building located in a southwestern city was vacated, and both its HVAC system and all furnishings in occupied spaces were replaced. In 1981, the building was reoccupied. NIOSH conducted follow-up environmental and epidemiological studies in this building.

Building D has 19 stories. The first 10 floors are occupied by offices. The open water spray system that was associated with the outbreak of HP in the middle 1970s was removed during building renovation. The new HVAC system contains 31 AHUs. Eleven of these units, one each on the basement through the 10th floor, provide conditioned air containing a variable amount of outside air mostly to central zones on each floor. The remaining 20 AHUs recirculate air in occupied spaces along exterior building walls.

Each central AHU contains a plenum wherein outdoor and return air are mixed, a roll filter without any rated atmospheric dust spot efficiency, a fan, and a bank of cooling and heating coils. Conditioned air leaving the central AHU is transported through ducts and delivered to occupied space on the same floor through ceiling diffusers. Air from occupied space enters the common plenum on each floor through grilles in the suspended ceiling. Return air moves either to peripheral AHUs for conditioning and recirculation to occupied space along peripheral floor zones or to the central AHU where return air is mixed with outdoor air prior to conditioning and transport back to central floor zones.

Stagnant water was found in drain pans located under cooling coils of central and peripheral AHUs in Building D. A growth of microbial slime several mm thick was found on wetted surfaces of drain pans and cooling coils. The access door to the cooling deck of each AHU was so tiny (460 cm^2 or 0.5 ft^2) that preventive maintenance was impossible for both the cooling deck and drain pan.

Air flow measurements made in the third floor central AHU operating under summer cooling conditions (minimum outdoor air dampers open) showed that only 262 L/s (525 cfm) of outdoor air was being taken into this unit. The number of employees on this floor was 93 indicating that an average of about 3 L/s (6 cfm) outdoor air (30% of ASHRAE minimal recommended level for occupied space where cigarette smoking is permitted)[18] was being provided per occupant. Employees working in zones served by peripheral AHUs are likely provided with amounts of outdoor air below this average value. Additional environmental measurements showed that occupied space was free of contaminant gases and that total suspended particulate in indoor air was below 50 $\mu g/m^3$.

Since the newly installed AHUs in Building D are contaminated with microbial slimes, the potential exists for a renewed outbreak of HP. Possible problems with microbial contamination are increased because of inadequate dilution by outside air. We recommended the installation of adequately sized access doors to the cooling deck portion of AHUs to assure and facilitate a preventive maintenance program to remove slime and stagnant water. We suggested that the HVAC system be operated according to AHSRAE guidelines.[18] The amount of outdoor air taken in by central AHUs might have to exceed the ASHRAE minimum of 10 L/s (20 cfm) per occupant so as to compensate for inadequate outdoor air in zones served by peripheral AHUs.

Building E

NIOSH conducted a health hazard evaluation in an office complex (located in an eastern city) housing more than 2000 employees. Some employees of an agency which occupied the lower floors of this building reported symptoms ranging from eye, nose, and throat irritation to persistent cough, shortness of breath, and fatigue. The building, constructed in 1969, has had a history of persistent indoor environmental problems including floods from roof leaks and a relative humidity that often exceeds 70% during the summer air conditioning season.

The central HVAC system of Building E contains 7 main AHUs that supply 100% outdoor air to occupied space. Conditioned air moves through common supply plenums formed by the suspended ceiling and the slab of the floor above. Supply air enters occupied space through diffusers located in the suspended ceiling. Air within Building E is further conditioned and recirculated by over 350 small AHUs and more than 1000 peripheral FCUs. Each small AHU is located in an interior zone, and it provides recirculated and conditioned air (cooling coils only) to occupants in several rooms through a small system of ducts. Peripheral FCUs are found in rooms along outside walls. These units condition (heat or cool) and recirculate air within occupied space. A more detailed description of the HVAC system of Building E is found elsewhere.[24] During several visits to this building it was observed that main AHUs were turned off for several hours during the working day.

Inspection of lower floors of Building E in January and September 1983, revealed evidence of moisture incursion into occupied space such as wet ceiling tiles and wet masonry. Tiles in some offices were partially covered by colonies of sporulating fungi. During the cooling season (September 1983) it was additionally observed that stag-

TABLE IV
Sequential Levels of Airborne Fungi in
Conference Room in Building E
Where Conditioned Air is Provided by a Small AHU

Temporal Sequence of Measurement	Operation of AHU[a]	CFU/m^{3b}
1	Unit off	240
2	Fan on during sampling	1650
3	Pound ductwork before sampling; fan on during sampling	1810
4	Ductwork pounded and fan on during sampling	3450
5	Filter replaced before sampling[c]; ductwork pounded and fan on during sampling	62,000
6	Pound filter during sampling; fan on	70,000

[a]Unit turned off between sequential sampling.
[b]Level of fungi in outdoor air = 800 CFU/m^3.
[c]Filter removed from unit two hours before sampling sequence initiated. Filter was heavily laden with dust and debris.

ductwork was pounded, and when the filter was replaced (Table IV). A seven-fold increase in levels of airborne fungi was associated with simply turning on the unit's fan. Pounding of ductwork during fan operation additionally doubled the level of airborne fungi. Fungal levels in the conference room increased more than an order of magnitude when the units's dirty filter was replaced and the ductwork was subsequently pounded.

Fungal levels present in the air of office rooms served primarily by FCUs were highly variable (Table V). In one room with an inactive FCU the concentration of airborne fungi was only 165 CFU/m^3; a nearby room with a characteristic "barnyard-like" odor contained over 7000 fungal CFU/m^3. The drain pan of the FCU in the latter room contained stagnant water. When the fan of this FCU was turned on and its outside metallic surface was agitated (similar to backing a chair into the unit) the level of airborne fungi became so high as to exceed the measuring capacity of the Andersen viable sampler ($>$ 94,000 CFU/m^3).

The generic compostion of airborne fungi isolated from selected rooms in Building E is given in Table VI. In the room with the contaminated FCU airborne *Penicillium* spp. accounted for more than 90% of isolates both with the unit off and on. In the conference room served by the small AHU *Cladosporium* spp. accounted for more than 80% of the isolates. In contrast, the outdoor air contained a more varied composition of flora. Al-

nant water and microbial slimes were present under cooling coils in some main AHUs and in some FCUs. Bacteria were present at a concentration of 1×10^7/ml in stagnant water found in one AHU. It could not be ascertained if microbial slime was present in drain pans of small AHUs because the cooling coils of each unit were totally inaccessible (sealed in room walls) for maintenance purposes.

Filters present in most AHUs and FCUs were heavily laden with dust and debris. Dust from a filter in a small AHU (January 1983) contained approximately 3×10^7 viable fungi/g. More than 90% of the isolates were *Penicillium* spp. All small AHUs and FCUs in Building E were lined along some interior surfaces with fibrous glass which was heavily encrusted with dust and debris. In some units fungal mycelia were found on insulation surfaces.

Sequential sampling for airborne fungi was carried out in an interior room where most of the conditioned air was provided by a small AHU. Air samples were collected on a table in the center of the room during various operating conditions including when the fan was turned on, when the

TABLE V
Levels of Airborne Fungi in Office in
Building E in Which Conditioned Air
is Provided by a Perimeter Fan Coil Unit

Room and Description of Fan Coil Unit	CFU/m^{3*}
Room #1 Had been occupied by complainant; unit off; stagnant water in drain pan	7,360
Room #1 Had been occupied by complainant; unit on and agitated; stagnant water in drain pan	$>$94,000
Room #2 Individual previously housed in room #1 now occupies this office without apparent complaint; unit off; no stagnant water in drain pan	165

* Level of fungi in outdoor air = 800 CFU/m^3.

TABLE VI
Generic Composition of Fungi in Building E and Outdoors*

Location	*Penicillium*	*Cladosporium*	*Alternaria*	*Aspergillus*	Other
Conference room; sequence 5 and 6 of Table IV	13.6	80.6	0	0.4	5.4
Room #1 Table V; unit off	>99	<0.1	0	<0.5	<0.1
Room #1 Table V; unit on	97.7	2.0	0	0	0.3
Outdoor air	12.5	48.6	12.5	0.7	25.7

* Percent of total airborne isolates in each location. Sampling and identification were carried out in September 1983.

though *Cladosporium* spp. accounted for almost half of the outdoor fungi, all air samples contained moderate numbers of *Penicillium* spp., *Alternaria* spp., and unidentified isolates.

It may be concluded that peripheral FCUs and small AHUs act as reservoirs for viable fungi in Building E. Because maintenance of mechanical systems in Building E is poor, events such as activating a FCU or AHU fan, changing filters, or even backing a chair into a peripheral FCU, may result in an elevated fungal level of up to 100× the outdoor concentration. The problem of high levels of airborne fungi in the occupied space of this building is additionally exacerbated by inadequate outdoor air ventilaton which makes dilution of viable aerosols less effective.

Recommendations for remedial action included the following:

1. Prevent moisture incursion into occupied space such as from drain pan overflows.
2. Replace filters of FCUs and AHUs routinely and frequently. Seal dirty filters in bags immediately after removal so as to prevent dissemination of spore clouds into occupied space. Replace dirty interior sound lining in these units.
3. Operate main AHUs during all times that the building is occupied.

Discussion

A common feature of office buildings wherein outbreaks of HP or of similar illnesses have occurred may be moisture incursion into occupied space or into the HVAC system. Thus Buildings A, B, and E were characterized by a history of repeated floods. In addition, all HVAC systems examined in this study contained mechanical systems with pools of stagnant water and deposits of microbial slimes. That moisture incursion can lead to elevated microbial levels which are associated with respiratory disease is evident in a residential case study of an individual with rhinitis.[25] Airborne sampling conducted subsequent to a roof leak revealed levels of viable fungi exceeding 5000 CFU/m^3. A level of only 260 CFU/m^3 was present before the flooding.[25] Since air is a ready source of microorganisms, and since substrate that supports microbial growth such as paper, plasterboard, ceiling tiles, carpet, and organic dusts are commonly found in office buildings, prevention of moisture incursion into occupied space and within HVAC systems is probably the best means of preventing microbially induced respiratory disease.

There is no environmental criteria for deciding if a measured airborne level of fungi or bacteria is a risk factor with regard to HP or other respiratory disease. Any quantitative criteria must take into account the qualitative nature of the viable and nonviable etiologic agents thought to be responsible for these illnesses. Problems associated with quantitative microbial standards have been previously discussed.[26] For example, is air containing a total of 500 fungi/m^3 of which 20% are *Penicillium* spp. inherently safer than that with 1000 fungi/m^3 but with only 10% *Penicillium* spp.? Establishment of a quantitative standard is further complicated because nonviable spores[25] and microbial products[6] may cause illness and large doses or organic dust may be needed to produce sensitization whereas a subsequent response may be evoked by a small quantity of material.[20]

Nevertheless, several suggestions have been made concerning acceptable levels of airborne viable particulate. In 1948 a level of approximately 1775 bacteria-containing particles/m^3 was described as the threshold for clerical offices in need of investigation and improvement.[27] Levels of about 700 bacteria/m^3 were considered reasonable. In 1969 it was stated that total levels of microorganisms exceeding 1700/m^3 were seldom exceeded in rooms.[28] During an outbreak of humidifier fever, levels of *Flavobacterium* spp. approximating 3000/m^3 were associated with the operation of a contaminated humidifier; this bacterium was absent from the air when the unit was turned off.[6] The threshold levels of *Cladosporium* spp. and *Alternaria* spp. spores for evoking allergenic symptoms have been reported to be 3000 and 100 CFU/m^3, respectively.[29] Levels of fungi approximating 5000 to 10,000 CFU/m^3 were associated with an outbreak of HP in a small office.[8] These literature citations collectively suggest that a level of viable microorganisms in excess of about 1×10^3 viable particles per m^3 indicates that the indoor environment may be in need of investigation and improvement. However, this is not to say that the air is unsafe or hazardous. Illness in the workplace can only be determined by medical or epidemiological studies.

Numerous and sometimes unpredictable variables affect results of aerobiological sampling carried out in office buildings. Among these are the following:

1. Sampling must occur in close temporal relationship to the event that triggers illness. In Building A, illness was associated with the activation of a HVAC system containing stagnant water, but unfortunately sampling occurred 10 days after this event. In Building B, sampling occurred at least a month after the floods that were related to illness.

2. Respiratory diseases such as HP are caused by a wide variety of microorganisms or microbial products that may require special sampling instrumentation[25,30] and special collection media.[7]

3. Sampling may be affected by the variable contamination and variable operational parameters of HVAC system components including AHUs, FCUs (Tables IV and V), humidifiers,[6] and water sprays.

4. Viable sampling indoors will be affected by sedimentation rate of airborne particles,[31] the effect of HVAC filtration units,[31] and the infiltration of seasonally varying loads and types of microorganisms in the outdoor air.[32]

While the above discussion indicates that microbial sampling is complex and affected by numerous variables, it has been shown to be of significant value in establishing disease etiology when closely integrated with medical and epidemiological investigations.[6-8,25]

It was earlier suggested that airborne microbial levels greater than 1×10^3/m^3 may be indicative of an indoor environment in need of improvement. Our studies have shown that two additional quantitative parameters, namely counts of microorganisms both in stagnant water and in dusts found in HVAC systems may be helpful in deciding if the indoor environment of a building is in need of improvement. Slime and stagnant water found in AHUs and FCUs in Buildings C and E contain microbial loads ranging from 1×10^5 to 1×10^7/ml. By contrast the reservoir of well maintained cooling towers is reported to contain only about 1×10^3 to 1×10^4 bacteria/ml.[33,34] While the microbial aerosols associated with cooling towers are generally found outside the building envelope, those aerosols arising from stagnant water and slimes within a HVAC system will likely be delivered directly to building occupants by the supply air stream. An additional parameter that may be useful in determining if the indoor environment is in need of improvement is the level of microorganisms present in dusts found in HVAC system components. Dust from filters of the type described in Table IV contains about 3×10^7 fungi/g. By contrast, house dust contains about 2×10^5 fungi/g.[35,36] Although much further study is necessary, our preliminary conclusion is that a level of bacteria or fungi in excess of 1×10^5/ml in stagnant water or slime and levels of fungi in excess of 1×10^6/g in dusts suggests that microbial contamination of HVAC system components is excessive.

According to ASHRAE Standard 62-1981[18] "indoor air should not contain contaminants that exceed concentrations known to impair health and cause discomfort to occupants." Microorganisms are among the contaminants listed by ASHRAE as possible indoor air pollutants. Unfortunately there are no studies of dose/response relationships that describe the concentration of microorganisms that may impair health or cause occupant discomfort. ASHRAE Standard 62-1981 does address the problems of certain occupant generated contaminants such as cigarette smoke and CO_2 by recommending that conditioned air contain a minimum of 10 and 2.5 L/s (20 and 5 cfm) of outdoor air/occupant for smoking and non-smoking environments, respectively. It is well known that humans shed microorganisms and that the number of bacteria in indoor air can be related to the number or density of room occupants.[37,38] Furthermore, airborne bacterial levels have been inversely related to room ventilation rates.[39] Dilution ventilation of the type recommended by ASHRAE appears to be adequate to lower airborne levels of microorganisms generated by occupants, at least when sources are moderate. In a study in a San Francisco office building the level of airborne bacteria was lowered from 179 to 105 CFU/m^3 when the supply of outdoor air per occupant was increased from 2-3 L/s (4-6 cfm) to 10-11.5 L/s (20-23 cfm).[40]

Not specifically addressed in ASHRAE 62-1981 is building-associated microbial contamination of the type described in our studies. This includes microorganisms found in HVAC system components such as water spray systems, humidifiers, AHU drain pans, and FCUs. Local exhaust ventilation as a remedial measure is impractical and the effects of increased dilution with outdoor air have not been studied.

Industrial hygiene preventive measures

Very little information is available on preventive and remedial measures that are effective in reducing building-associated microbial contamination. Future studies in this area are urgently needed. Listed below are some preventive measures which in our professional judgement may be effective in reducing building-associated microbial contamination and building-associated HP illnesses. As such, it would be prudent for building engineers and industrial hygienists to incorporate at least some of the following into their preventive maintenance programs.

1. Repair all external and internal leaks promptly and permanently.

2. Stagnant water should not be allowed to accumulate under cooling coils in any type of AHU. Proper inclination and continuous drainage of drain pans is required. Cooling coils should be run at a low enough temperature so that adequate dehumidification can result to keep space relative humidities at proper levels and so that spores and substrate impacted on coils may be washed away in condensate water. AHU components should be inspected for the presence of slime and stagnant water. AHUs must be constructed or modified so that maintenance personnel have easy and direct access to the heat exchange components as well as to drain pans. If contamination with microbial slime is found, it must be removed. Mechanical or detergent cleaning may be necessary to remove slime before using microbiocidal chemicals. Steam lancing can be used to remove slime providing that the treatment does not damage heat exchange surfaces.[41] Chlorine generating slimicides and proprietary biocides may be used for disinfection provided that these chemicals are removed before AHUs are reactivated.[41] Aerosolization of microbiocidal chemicals into occupied space must never occur.

3. Humidifiers in HVAC systems should preferentially use steam as a moisture source. Raw stream from the plant boiler system generally contains corrosion inhibitors which are meant to carry over into condensate return lines. Therefore, it may be more desirable to use a small separate steam generator for humidification purposes. Also in many instances large boilers are not operable year around. Humidifiers utilizing recirculated water are not recommended as these almost always become rapidly contaminated with organic dusts and microorganisms. The addition of biocides to this type of humidifier has been ineffective in controlling microbial contamination.[5,42] If cold water type humidifiers are used, water should originate from a potable source and water should be run to a drain[7] instead of being

recirculated. Cold water humidifiers should also be subject to a fastidious preventive maintenance program[43] involving regular inspection, cleaning, and disinfection as outlined above in point two. The use of portable cold mist vaporizers is discouraged since these devices are known to readily contaminate air with microorganisms.[44]

5. Relative humidity in occupied spaces and low air velocity plenums should not exceed 70%. As relative humidity rises above this level, the increased moisture content of organic substances encourage fungal spore germination and proliferation.[35,46] This recommendation is contrary to generally accepted energy management techniques wherein occupied space relative humidities can exceed 70% during the cooling months. Relative humidity can be lowered either by reducing the moisture content or raising the temperature of the air. Cooling coils of AHUs must be run at a low enough temperature to dehumidify conditioned air. In buildings using economizer systems where relative humidity is sometimes excessive, outdoor air may require dehumidification. The sensible heat (dry bulb temperature) of the outdoor air may be suitable but the total enthalpy may be unacceptable. Reheat coils may have to be utilized to raise the temperature of dehumidified supply air as the dry bulb temperature of this air may be too low due to the prior necessary dehumidification process.

6. Filters used in AHUs should have a moderate (50-70%) efficiency as measured by the ASHRAE atmospheric dust spot test[21,47] and in general should be the extended surface type. To prolong the life of these filters and to improve cost effectiveness, prefilters (such as roll type) should be used to clean the air prior to passage over the higher efficiency filters. Filters of this efficiency will remove spores as well as organic dusts that support microbial growth. Electronic air cleaners have also been reported to be effective in removing microorganisms from the air stream,[48,49] but first cost and maintenance may prohibit their use. Care also must be exercised to insure that these air cleaners not liberate ozone into the airstream. The location of filters in the HVAC system is important. In most AHUs, filters are located upstream of the heat exchange components so as to protect the heat exchange capacity of these surfaces. Building occupants will not be entirely protected by these conventionally located filters if microbial contamination occurs downstream (e.g., slime in cooling coil drain pan). Therefore, it is sometimes necessary to provide filtration downstream from heat exchange components to achieve suitable protection. Because dirty filters in AHUs are sources of microorganisms (Table IV)[8] a preventive maintenance program must exist by which filters are inspected and replaced at regular intervals.

7. Microbial aerosols from cooling tower drift, sanitary, and other exhaust vents may enter improperly located outdoor air intakes. Remedial action could include relocation of the vents or increasing their heights appropriately, relocation of outdoor air intakes, or upgrading AHU filter efficiency. A preventive maintenance program to inhibit slime build-up in cooling towers is essential.[50,51] Outdoor microbial aerosols may also enter buildings through idle exhaust ducts and miscellaneous stacks and vents.[52] For this reason, the buildiing should be operated so that inside pressure is slightly positive (5 to 8 mm water gauge or 0.02 to 0.03 inches water gauge)[53] with respect to the outdoors.

8. In buildings or zones of buildings where carpet, upholstery, ceiling tiles, and other porous furnishings are contaminated with microbial material, it is better to discard these items rather than to attempt disinfection (Building B).[25] Since it is almost impossible to clean contaminated suspended ceiling-return air plenums,[38] these may have to be bypassed by installation of return air ductwork. A special maintenance program may be needed to clean FCUs. This should include the removal of debris from beneath and within units[54] and replacement of dirty insulation along inside surfaces. During building clean-up, microbially laden materials (e.g., carpets, filters, etc.) should be carefully removed so as to minimize aerosolization of inhalable particulate. Personnel assigned to clean-up operations should at a minimum wear respirators with high efficiency particulate filter media. Structural and other building surfaces should be vacuumed with an instrument

incorporating a HEPA filter and then disinfected with bleach or proprietary biocide.

9. At a minimum, outdoor air should be provided in conditioned air at a rate of 10 and 2.5 L/s (20 and 5 cfm) per occupant for smoking and non-smoking environments, respectively.[18] Higher levels of outdoor air intake may be required to effectively dilute microbial aerosols associated with contaminated buildings.

References

1. Weiss, N.S. and Y. Soleymani: Hypersensitivity Lung Disease Caused by Contamination of an Air Conditioning System. *Ann. Allergy 29*:154-156 (1971).
2. Banaszak, E.F., W.H. Thide and J.N. Fink: Hypersensitivity Pneumonitis due to Contamination of an Air Conditioner. *N. Eng. J. Med. 283*:271-276 (1970).
3. Scully, R.E., J.J. Galdabini and B.U. McNeely: Case Records of the Massachusetts General Hospital, Case 47-1979. *Ibid. 301*:1168-1174 (1979).
4. Arnow, P., J. Fink, D. Schlueter et al: Early Detection of Hypersensitivity Pneumonitis in Office Workers. *Am. J. Med. 64*:236-242 (1978).
5. Ganier, M., P. Lieberman, J. Fink and D.G. Lockwood: Humidifier Lung: An Outbreak in Office Workers. *Chest 77*:183-187 (1980).
6. Rylander, R., P. Haglind, M. Lundholm et al: Humidifier Fever and Endotoxin Exposure. *Clin. Allergy 8*:511-516 (1978).
7. Edwards, J.H.: Microbial and Immunological Investigations and Remedial Action after an Outbreak of Humidifier Fever. *Brit. J. Ind. Med. 37*:55-62 (1980).
8. Bernstein, R.S., W.G. Sorenson, D. Garabrant et al: Exposures to Respirable, Airborne *Penicillium* from a Contaminated Ventilation System: Clinical, Environmental and Epidemiological Aspects. *Am. Ind. Hyg. Assoc. J. 44*:161-169 (1983).
9. Andersen, A.A.: New Sampler for the Collection, Sizing, and Enumeration of Viable Airborne Particles. *J. Bacteriol. 76*:471-484 (1958).
10. Morring, K.L., W. Jones, P. Morey and W. Sorenson: Evaluation of a Single Stage Viable Sampler. *Am. Ind. Hyg. Conf.* (1984).
11. Lippmann, M.M. and T.L. Chan: Calibration of Dual Inlet Cyclones for "Respirable" Mass Sampling. *Am. Ind. Hyg. Assoc. J. 34*:189-200 (1974).
12. U.S. Dept. Health, Education, and Welfare: Criteria for a Recommended Standard — Occupational Exposure to Fibrous Glass. NIOSH Pub. No. 77-152 (1977).
13. Johnson, C.L., I.L. Bernstein, J.A. Gallagher et al: Familial Hypersensitivity Pneumonitis Induced by *Bacillus substilis*. *Am. Rev. Resp. Dis. 122*:339-348 (1980).
14. Patterson,, R., J.N. Fink, W.R. Miles et al: Hypersensitivity Lung Disease Presumptively Due to *Cephalosporium* in Homes Contaminated by Sewage Flooding or by Humidifier Water. *J. Allergy Clin. Immunol. 68*:128-132 (1981).
15. Wolf, F.T.: Observations on an Outbreak of Pulmonary Aspergillosis. *Mycopath. et Mycol. Appl. 38*:359-361 (1961).
16. Solley, G.O. and R.E. Hyatt: Hypersensitivity Pneumonitis Induced by *Penicillium* species. *J. Allergy Clin. Immunol. 65*:65-70 (1980).
17. Hodgson, M.J., P.R. Morey, M. Attfield et al: Pulmonary Disease Associated with Cafeteria Flooding: Single Breath Carbon Monoxide Diffusing Capacity in a Cross-sectional Field Study. *Arch. Env. Health* (submitted).
18. American Society Heating, Refrigerating, and Air-Conditioning Engineers: *Standard 62-1981: Ventilation for Acceptable Indoor Air Quality*. Atlanta, GA (1981).
19. U.S. Dept. Health, Education, and Welfare: *Criteria for a Recommended Standard — Occupational Exposure to Carbon Dioxide*. NIOSH Pub No. 76-194 (1976).
20. Medical Research Council Symposium: Humidifier Fever. *Thorax 32*:653-663 (1977).
21. American Society of Heating, Refrigerating, and Air-Conditioning Engineers: *Standard 52-76: Method of Testing Air-Cleaning Devices Used in General Ventilation for Removing Particulate Matter*. Atlanta, GA (1976).
22. Miller, R.P.: Cooling Towers and Evaporative Condensers. *Ann. Int. Med. 90*:667-670 (1979).
23. American Industrial Hygiene Association: *Heating and Cooling for Man in Industry*, 2nd ed., pp. 61-64. Akron, OH (1975).
24. Morey, P.R.: *Case Presentation: Problems Caused by Moisture in Occupied Spaces of Office Buildings*. (In this volume.)
25. Kozak, P.P., Jr., J. Gallup, L.H. Cummins and S.A. Gillman: Currently Available Methods for Home Mold Surveys. II. Examples of problem homes surveyed. *Ann. Allergy 45*:167-176 (1980).
26. Greene. V.W., D. Vesley, R.G. Bond and G.S. Michaelsen: Microbiological Contamination of Hospital Air. II. Qualitative studies. *Appl. Microbiol. 10*:567-571 (1962).
27. Bourdillon, R.B., O.M. Lidwell, J.E. Lovelock and W.F. Raymond: Airborne Bacteria Found in Factories and Other Places: Suggested Limits of Bacterial Contamination. *Studies in Air Hygiene*, pp. 257-263. Medical Res. Coun. Spec. Rep. Series No. 262. HM Stationery Office, London (1948).
28. Wright, T.J., V.W. Greene and H.J. Paulus: Viable Microorganisms in an Urban Atmosphere. *J. Air. Poll. Cont. Assn. 19*:337-341 (1969).
29. Gravesen, S.: Fungi as a Cause of Allergenic Disease. *Allergy 34*:135-154 (1979).
30. Reed, C.E., M.C. Swanson, M. Lopez et al: Measurement of IgG Antibody and Airborne Antigen to Control an Industrial Outbreak of Hypersensitivity Pneumonitis. *J. Occup. Med. 25*:207-210 (1983).
31. McNall, P.E.: Practical Methods of Reducing Airborne Contaminants in Interior Spaces. *Arch. Env. Health 30*:552-556 (1975).
32. Solomon, W.R.: Assessing Fungus Prevalence in Domestic Interiors. *J. Allergy Clin. Immunol. 56*:235-242 (1975).
33. Adams, A.P. and B.G. Lewis: Microbiology of Spray Drift from Cooling Towers and Cooling Canals. *Proc. Conf. Microbiology of Power Plant Effluent*, pp. 31-36. Univ. of Iowa Press, Iowa City (1978).
34. DeMonbrun, J.R.: Personal Communication. Union Carbide Corporation, Oak Ridge, TN.
35. Brundrett, G.W. and A.H.S. Onions: Molds in the Home. *J. Consumer Stud. Home Econ. 4*:311-321 (1980).

36. **Swaebly, M.A. and C.M. Christensen:** Molds in House Dust, Furniture Stuffing, and in the Air Within Homes. *J. Allergy* 23:370-374 (1952).
37. **Dimmick, R.L. and W. Wolochow:** Effects of Energy Conservation Measures on Air Hygiene in Public Buildings. *Final Report for U.S. Dept. of Energy,* under Contract No. W-7405-ENG-48 (1980).
38. **Gundermann, K.O.:** Spread of Microorganisms by Air Conditioning Systems — Especially in Hospitals. *Ann. N.Y. Acad. Sci.* 353:209-217 (1980).
39. **Williams, R.E.O., O.M. Lidwell and A. Hirsh:** The Bacterial Flora of the Air of Occupied Rooms. *J. Hyg.* 54:512-523 (1956).
40. **Turiel, I., C.D. Hollowell, R.R. Miksch et al:** The Effects of Reduced Ventilation on Indoor Air Quality in an Office Building. *Atmos. Env.* 17:51-64 (1983).
41. **Brundrett, G.W., J.B. Collins, G.S. DaCosta et al:** Humidifier Fever. *J. Chart. Inst. Build. Serv.* 3:35-36 (1981).
42. **Cockcroft, A., J. Edwards, C. Bevan et al:** An Investigation of Operating Theatre Staff Exposed to Humidifier Fever Antigens. *Brit. J. Ind. Med.* 38:144-152 (1981).
43. **Brundrett, G.W.:** *Maintenance of Spray Humidifiers.* The Electricity Council, Research Center, Capenhurst, Chester, England (1979).
44. **Solomon, W.R.:** Fungus Aerosols Arising from Cold-mist Vaporizers. *J. Allergy Clin. Immunol.* 54:222-228 (1974).
45. **Bynum, L.M.:** Personal communication. Monsanto Textiles Company, Pensacola, FL.
46. **Block, S.S.:** Humidity Requirements for Mold Growth. *Appl. Microbiol.* 1:287-293 (1953).
47. **Barnstable, D.:** Basic Considerations for Selecting Air Filters. *Plant Engin.* 45:145-147 (1981).
48. **Margard, W.L. and R.F. Lodgsdon:** An Evaluation of the Bacterial Filtering Efficiency of Air Filters in the Removal and Destruction of Airborne Bacteria. *ASHRAE J.* 10:49-54 (1965).
49. **Kozak, P.P. Jr., J. Gallup, L.H. Cummins and S.A. Gillman:** Factors of Importance in Determining the Prevalence of Indoor Molds. *Ann. Allergy* 43:88-94 (1979).
50. **Sibley, H.:** Legionnaires' Disease: A Status Report. *ASHRAE J.* 23:56-57 (1981).
51. **American Society of Heating, Refrigerating, and Air-Conditioning Engineers:** *Systems Handbook,* pp. 36.14-15. Atlanta (1980).
52. **Band, J.D., M. LaVenture, J.P. Davis et al:** Epidemic Legionnaires' Disease, Airborne Transmission Down a Chimney. *JAMA* 245:2404-2407 (1981).
53. **Morris, R.H.:** Personal communication. Air Monitor Corporation, Parsippany, NJ.
54. **Shaffer, J.G.:** Air-Borne *Staphylococcus aureus,* A Possible Source in Control Equipment. *Arch. Env. Health* 5:547-551 (1962).

Epidemiologic investigation of office environmental problems

DEAN B. BAKER, M.D., MPH
School of Public Health, University of California — Los Angeles

The prevalence of health complaints among persons working in modern office buildings has increased dramatically during the past few years. These complaints range from clearly defined respiratory infections to non-specific symptom patterns, including headache, fatigue, and mucous membrane irritation. Public health officials, as well as private industrial hygienists and ventilation engineers, have been asked to determine the cause of these health problems. The diversity and persistence of complaints suggest that the etiology of the health problems is multifactorial and that effective solutions will require a team approach. This paper will discuss the role and methods of medical epidemiology in the evaluation of office environmental health problems.

The role of epidemiology in evaluating office environments has increased substantially during the past few years as the nature of the problem has evolved. Initially, most health complaints occurred in fairly specific patterns which implicated particular problems. For example, occupants may have reported developing headaches and smelling car exhaust following the construction of a new parking garage next to the building's ventilation system intake. The investigator could identify the problem by interviewing a few of the building's occupants during an initial visit and then confirm the problem by using appropriate environmental sampling techniques. Epidemiologic methods were not necessary to solve these problems.

During the past few years, the proportion of health complaints due to easily identifiable causes has decreased. Today public health investigators are faced with a wide range of health complaints among building occupants. These occupants also may report experiencing odors, stuffy air, and thermal discomfort. Environmental sampling usually reveals no specific agent, although there may be dozens of chemicals present at extremely low air concentrations. The specific etiology of the complaints is not found and the investigator only can suggest that the ventilation rate be increased and hope that the complaints decrease.

Epidemiology — the study of disease in populations — can be useful to the investigator in evaluating these modern office problems by identifying patterns which suggest new hypotheses and lead to appropriate environmental sampling strategies. Epidemiology also can be used to assess social factors and health outcomes not within the traditional purview of industrial hygiene techniques. Finally, epidemiological and statistical analysis may be required to estimate the relative contributions of multiple chemical, physical, and social factors in causing the modern, complex office environmental problems.

The fundamental premise of epidemiology is that disease does not occur randomly in the population; rather, it occurs in patterns which are determined by the characteristics of the agent, host (or affected person), and environment. Epidemiologists attempt to discover the cause of health problems by systematically observing patterns in the occurrence of the disease. Since pattern recognition requires some understanding of possible patterns, I will review the major agents and health complaints before discussing the epidemiologic approach to evaluating these problems.

Agents

Agents may be classified as being "Biological," "Chemical," "Physical," or "Social." Evaluation methods are quite different for these classes of agents, but their health effects overlap substantially. For example, a headache could be due to any of these classes of agents. The investigator should look for patterns of health complaints which suggest specific agents or classes of agents which then can be further evaluated.

Biological agents

Biological agents, such as bacteria, viruses, rickettsiae, and fungi, rarely cause office environmental problems; however, the illnesses they do cause can be dramatic. These agents affect health by causing infections or by acting as allergens. The health symptoms and patterns of occurrence due to these two mechanisms are quite different.

Building-related infections are due to a limited number of pathogenic agents which flourish within the building's ventilation system and then spread to susceptible occupants. Infections which spread from person to person within a building generally are not considered as being due to the building. For example, influenza is not considered a "building-related" infection even though it is likely to spread among building inhabitants. The building is not intrinsic to the chain of transmission of the infection. On the other hand, the building does play an integral role in the transmission of some noncontagious infections, such as Legionnaires Disease. The humidification system of the building acts as a reservoir where the agent grows, while the ventilation system transmits the agent to susceptible persons within the building. These "building-related" infections can be recognized by their relatively specific health effects, such as pneumonia, and by their epidemic occurrence among the building inhabitants.

Some biological agents affect health by causing allergic reactions — allergic rhinitis (hayfever), contact dermatitis, asthma, and hypersensitivity pneumonitis. Most allergic reactions are idiosyncratic in that any individual can develop an allergic reaction to virtually any agent. Thus, there is no clear pattern of occurrence and building occupants may not even associate their symptoms with the building. However, certain bacteria and fungi have been implicated in epidemics of hypersensitivity pneumonitis attacking up to 70% of exposed occupants.[1] Because biological agents can cause such a wide range of health effects, the investigator must consider them in many instances where a direct toxic effect has been ruled out.

Chemical agents

Thousands of chemical substances — gases, vapors, and particulates — are present in the modern office environment. They come into the building from outside, they derive from the fabric of the building, and they are generated from the activities of the building's inhabitants. Most chemicals can cause adverse health effects if present at a sufficient dose. Some substances cause specific definable health effects; however, many are capable of causing the more general effects of mucous membrane irritation and/or central nervous system depression, resulting in headache, dizziness, and fatigue. The effects of multiple chemical agents in causing these general health complaints are additive, if not synergistic.

It is difficult to interpret the significance of finding an extensive mixture of gases and vapors, with each substance being present at a very low air concentration. An example of typical environmental sampling results is shown in Table I. In this investigation of a New York office building, the investigators tabulated the results of environmental sampling by two industrial hygiene consulting firms and three governmental agencies over an eight month period.[2] There were dozens of chemicals present, but even at the highest concentrations detected, each was substantially below current recommended standards. Nevertheless, the non-specific health complaints of the workers in the building steadily increased until the office had to be abandoned. Most investigators now agree that exposure guidelines based on industrial settings may not be appropriate for office environments. However, better guidelines have not been developed yet.

Like biological agents, some chemicals may act as allergens. Only a small proportion of the inhabitants develop the allergy, but the ones that do can develop severe reactions with minimal exposure to the chemical. The few number of persons affected and the severity of the reaction help one differentiate these allergic reactions from more common, but milder irritation due to the chemical.

For several years, chemical agents have been the traditional purview of industrial hygienists involved in evaluating office environmental problems. The results of these investigations have been discouraging in that specific agents rarely are found. As exemplified by the topics of this symposium, industrial hygienists have begun to look to physical factors, such as inadequate ventilation, to provide an explanation for the health complaints reported by office building inhabitants.

Physical agents

The major physical agents include temperature, humidity, air flow, noise, light, and more generally the ergonomic design of the office space. Except in

TABLE I
Environmental Contaminants in a New York Office Building[2]

Substance	Lowest Recommended Level* ppm	Lowest Recommended Level* mg/m³	Highest Level Found
Carbon monoxide	35	—	3.0 ppm
Nitric oxide	25	—	0.16 ppm
Nitrogen dioxide	1	—	0.1 ppm
Ozone	0.1	—	0.05 ppm**
Ammonia	50	—	5.0 ppm**
Cellosolve	200	—	0.01 ppm
Epichlorohydrin	8	—	0.04 ppm
Formaldehyde	0.8	—	0.07 ppm
Freon 113	1000	—	0.7 ppm
"Hydrocarbons"	—	—	1.8 ppm
Hexane	100	—	Trace
MEK	200	—	Trace
Toluene	100	—	Trace
Trimellitic anhydride	—	0.05	Ceiling Bulk
Carbon black	—	3.5	Trace
Rock Wool	—	10.0	Trace
Fibrous glass	3 f/cc		Trace

* Based on then current OSHA Standards, NIOSH Recommended Standards, or ACGIH Threshold Limit Values.

** Limit of detection.

extreme circumstances, physical agents do not represent a direct health threat to office building inhabitants. They can contribute to discomfort and thus exacerbate symptoms and health concerns. Several investigators have reported that on questionnaires administered to occupants of "problem" buildings, health complaints were most strongly associated with perceptions of "inadequate ventilation," "stuffy air," and workstation discomfort.[3] In most instances, the ventilation in these buildings was not measured, but the investigators concluded that inadequate ventilation was the basis of the problem. It is important that investigators actually measure ventilation rates before they conclude that inadequate ventilation is reponsible for the health complaints.

The mechanisms by which perceptions of inadequate ventilation cause ill-health are unclear. Inadequate ventilation may allow the accumulation of toxic substances, but such accumulations have not been found. Investigators should consider the physiological mechanisms by which inadequate ventilation may reasonably explain the reported health complaints in the absence of known toxic exposures. It is likely that physical factors play an important role in causing office environmental problems; however, that role is far from clear. Most likely, physical factors, such as inadequate ventilation, are only one facet of the multifactorial etiology of health complaints. Until more information is available, it is important for industrial hygienists not to move all their "eggs" from the basket of "toxic chemicals" to the basket of "inadequate ventilation."

Social factors

With the focus initially on chemical agents and more recently on physical agents, social factors have been ignored. In fact, these factors could be key to understanding the dynamics of health complaints in those situations where there is no clear pattern of occurrence nor specific toxic agent present. It is worthwhile to view the modern office building as a semi-isolated ecologic system in which the building inhabitants carry out defined tasks. As such, the system has a social, as well as physical and chemical, fabric. Key elements include the structure of the occupant organizations, individual job stressors, and social networks among the workers.

Occupational stress research over the past decade has demonstrated the importance of these social factors in affecting health.[4] In fact, many of the health complaints of office building inhabitants parallel the manifestations of stress reported in the stress literature. The impact of these factors within the office environment has not been adequately assessed to date. Epidemiologic methods of questionnaire administration and analysis could be particularly valuable in this area.

It should be noted that social factors impact on office problems in two ways. First, organizational structure and work-task characteristics may cause or contribute to health problems by acting as psycho-social stressors. As mentioned above, headache can be a manifestation of stress. In addition, social factors are integral to the inhabitants' awareness of the office environmental problem. Virtually every investigator who has been involved in evaluating office environmental problems is aware that there usually is a clear social dynamic to the development of the complaints among the workers. This social dynamic is the organization's response to the environmental perturbation. Assessment of the organizational factors and social networks which are involved in this social dynamic is valuable in the development of effective intervention strategies.

Health problems

In most instances, an investigator is asked to evaluate a building after the inhabitants have perceived that there is a health problem. These problems range from easily definable illnesses through a melange of non-specific symptoms. The investigator first should attempt to identify specific illnesses or symptom patterns which are indicative of particular agents. Examples of specific illnesses include respiratory tract infections such as Legionnaires Disease and allergic reactions like hypersensitivity pneumonitis. The recognition and evaluation of these problems have been discussed elsewhere in this symposium.

Dermatitis

Dermatitis is another clearly apparent health problem; however, the etiologic agent can be surprisingly elusive. Contact dermatitis can be due to irritation or allergy. Allergic contact dermatitis looks no different than irritant contact dermatitis, but it affects only a small proportion of exposed workers. Most irritating substances are not present in sufficient quantities in offices to cause significant dermatitis; however, exposed persons may complain of itching and have an evanescent rash.

Several investigators have observed an irritant contact dermatitis due to fibrous glass or mineral wool insulation.[1] Diagnosis of the problem was made by the typical description and appearance of the rash, and in some instances, by microscopic examination of skin scrapings. Air concentrations of the fibers did not correlate with the development of the rash, probably because random environmental samples can not detect the intermittent settling of fibers from the ceilings and ventilation system onto the desk tops. On the other hand, occupants may associate the onset of their rashes with the installation of new insulation or with maintenance workers removing ceiling panels. Questionnaires to systematically assess occupants' recall of events can be quite useful in solving these problems.

Non-specific health problems

Unfortunately (for the investigator), the great majority of health problems are not as clear as respiratory infections nor as obvious as contact dermatitis. The most common health complaints among building inhabitants are symptoms such as headache, fatigue, and mucous membrane irritation (e.g., irritation of the eyes, nose, mouth, and throat). In many instances, the prevalence of these complaints among building inhabitants can be greater than 50%. Results of a typical symptom questionnaire are shown in Tables II. The prevalence of complaints is high, but the pattern is not indicative of exposure to one or a few toxic agents.

One may refer to these symptoms as being "Non-specific" because they can be due to many agents and their presence does not suggest a particular etiology. Non-specific does not mean that these complaints are unreal nor that they are due to hysteria. The etiology of these complaints is complex and most likely involves the interaction of chemical, physical, and social factors. A major role for epidemiology is in the evaluation of the multifactorial etiology of these non-specific health complaints

Epidemiologic evaluation

The major goals of any environmental health investigation are to determine whether the environment is safe or not; to discover the chain of events leading to the reported health problem; and to identify intervention strategies to alleviate the existing problem and prevent future problems. The epidemiologist complements the activities of the industrial hygienist and ventilation engineer in achieving these goals. Since most investigations are initiated due to health complaints, the epidemiologist can assess the nature and extent of the health problem. This assessment should generate hypotheses which provide direction to environmental sampling strategies. Epidemiologic techniques should be used to confirm health effects through appropriate medical testing and to evaluate physical and social factors which can not be assessed by industrial hygiene methods. Finally, epidemiologic methods can be used to analyze the multifactorial etiology of health complaints when no single agent is apparent.

The major steps of an epidemiologic evaluation are shown in Table III. During the initial visit, the investigator should attempt to establish the existence and extent of the health problem. It is important to elicit the views of the building inhabitants so that the investigation will be responsive to their concerns. At the same time, the investigator should review medical records and contact appropriate medical personnel to verify reported medical diagnoses. An epidemic, by definition, is an unexpectedly high incidence of disease in a population; thus, the epidemiologist must estimate both the number of people affected (the numerator of the incidence rate) and the number of persons at risk (the denominator) to tell whether an epidemic exists. A common pitfall is for the investigator to rely on a few vocal persons to "tell the story." The impressions of these self-appointed spokespersons may not be reliable. The epidemiologist must use all available information sources to derive estimates of the rates of health problems in order to define the extent of the problem.

A primary objective of this early data gathering is to describe the pattern of the building problem in

TABLE II
Frequency of Health Symptoms Among 285 Office Employees*[6]

Health Symptoms**	Percent of Respondents Reporting Having the Symptom		
	(Never or Hardly Ever)	(Some of the Time)	(At Least a Large Part of the Time)
Dry skin	32.6	27.7	39.6
Dry or scratchy throat	31.9	46.7	21.4
Nasal congestion	41.1	38.2	20.7
Extreme fatigue at work	36.8	43.2	20.0
Eye irritation (tearing)	50.5	33.0	16.5
Back pain	50.9	35.1	14.0
Severe headache	46.0	42.5	11.6
Colds or respiratory Infections	47.0	41.4	11.6

* Responses to the question: "How often have you experienced each of the following symptoms within the past year?"

** Symptoms reported by less than 10% of respondents to occur at least a large part of the time included: dizziness, nausea at work, blurred vision at work, persistent cough, trouble breathing, numbness or tingling in part of the body, and skin irritation or rash.

TABLE III
Epidemiologic Evaluation of Building-Related Problems

1. Establish existence and extent of problem.
    a. Orientation and walkthrough.
    b. Case verification.
2. Look for patterns by TIME, PLACE, PERSON.
    a. Rule out specific exposures.
    b. Any recent CHANGES?
3. Formulate hypotheses to direct further study.
    a. Environmental monitoring.
    b. Medical testing
4. Systematically evaluate occupants.
    a. Random (stratified) sample.
    b. Use COMPARISON Group, if possible.
    c. Assess health complaints, social factors, perceptions of environment.
5. Analyze multifactorial problem.
6. Present findings.
    a. Written report and presentation.
    b. Recommend interventions.

---

terms of "Time," "Place," and "Person." Specific patterns may become apparent which suggest particular etiologic agents. A key question in ruling out specific exposures is asking whether there had been any recent changes. Sometimes the chain of events can be quite complicated. For example, workers in the data processing center of a New York area hospital complained of smelling odors and developing headaches intermittently during the workday.[5] The investigation revealed that the ventilation system intake for the office was near a parking garage next to the hospital. The health complaints developed after the city changed street parking regulations which resulted in increased usage of the parking garage. At shift changes, many hospital employees were forced to sit in their idling cars waiting to exit the garage. The exhaust from the idling cars was drawn into the ventilation system, resulting in the complaints of the data processing personnel. The onset and timing of the complaints led to the understanding of the etiology of the problem. Baffles were placed to separate the ventilation system intake from the garage and the complaints resolved.

Unfortunately, many times the etiology of the problem will not be apparent after the initial walkthrough and investigation. In these instances, the detailed description of the problem by time, person, and place should lead to new hypotheses which can be verified by environmental monitoring and specific medical testing. Throughout the investigation, the epidemiologist and industrial hygienist work together as a team to continually refine the definition of the problem, which in turn will generate new hypotheses to be tested.

In parallel with the environmental assessment, the epidemiologist should conduct a systematic evaluation of building occupants. This assessment should be able to confirm initial impressions, as well as to detect subtle patterns which were not apparent at first. It is critical that the epidemiologist determine which persons will participate in the survey. The sampling of study subjects should be random or random among stratified groups (for example, certain floors of a building). The purpose of random sampling is to insure that participants in the survey are representative of the whole population at risk. Volunteers and persons with the most health complaints usually are not typical and should not be given particular weight when selecting the study participants.

Whenever possible, the investigator should recruit a comparison group to participate in the study. The purpose of studying a comparison group is to provide an estimate of the inherent or underlying rate or (risk) of disease in an unexposed population. By comparing the experiences of the two groups, the investigators may be able to determine why the study group has a higher rate of disease.

TABLE IV
Factors to be Assessed in an Epidemiologic Survey

1. Office Environment
    a. Temperature, humidity, air flow.
    b. Noise, light, odors.
    c. Equipment, furnishings.
2. Work Station
    a. Flexibility, comfort.
    b. Crowding, privacy.
3. Task Characteristics
    a. Demands, deadlines, role conflict.
    b. Discretion, repetition, monotony.
4. Organizational Chacteristics
    a. Hierarchial structure, policy input.
    b. Flexibility, communication.
5. Personal Factors
    a. Age, sex, ethnicity.
    b. Smoking, atopic status, medical history.

Comparison groups may be drawn from many sources. They may be non-affected persons within the same office, persons on different floors (particularly if supplied by a different ventilation unit), or even persons in a nearby building. Because an "unexposed" comparison group must be truly unexposed, the epidemiologist and industrial hygienist should work together to decide an appropriate sampling scheme based on their knowledge of potential environmental exposures.

Factors to be assessed

During the systematic evaluation, the epidemiologist should assess the subjects' perceptions of the office environment; social factors which may affect the health problem; and personal factors which may bias the reports of study subjects. These factors are summarized in Table IV.

Physical factors within the office environment — such as temperature, humidity, air flow, noise, and lighting — are much more likely to cause discomfort that direct toxic effects. Thus, it is relevant to assess the perceptions of the building inhabitants. These perceptions then can be linked to objective measures of the physical factors. In addition, it should be noted that environmental monitoring usually occurs during a brief period of time which may not be representative of typical conditions. Occupants should be asked to report their perceptions of typical building conditions.

Ergonomic features of the work station — such as flexibility, comfort, crowding, and privacy — are difficult to measure objectively, but may have a substantial impact on the comfort of office workers. Perceptions of these factors should be included on a survey questionnaire. Occupational stressors due to the work task characteristics and the organizational structure also should be assessed. Questions on these factors can be adapted from the stress research literature. Specific questions relating to the social dynamics of the problem also should be included. For example, it is important to understand how individual workers became aware that the building has an "environmental problem" and to assess whether this awareness may have affected their reporting of health problems.

Finally, the investigation should assess personal factors — such as age, sex, ethnicity, and smoking habits — which may influence susceptibility or the responses of study participants. The comparison between two study groups (e.g., exposed and unexposed) may be biased if the two groups are substantially different in these personal factors which may influence the incidence of health problems. For example, if the comparison group consists of younger persons than the study population and if age were associated with complaints of headaches, then the crude comparison of the two groups could be biased due to the difference in ages between the groups. A major task of epidemiology is to anticipate and avoid (or at least adjust for) these potential biases.

Analysis of results

After the investigators have completed their measurements of the office environment and inhabitants, they can use epidemiologic methods to analyze the results. The major objective of the analysis is to identify significant associations between exposure to specific agents and the incidence of health problems. As mentioned above, these associations are tested by comparing exposed groups of subjects with unexposed groups or by comparing diseased groups with similar non-diseased (control) groups.

The validity of the analysis is dependent on the investigators being able to recognize and adjust for potential biases in the data. Such biases may arise from differences between the study group and the comparison group (called confounding), from faulty recall of subjects, and from misclassification of exposure status by the investigators. A high quality epidemiologic study should include a conscious assessment of the potential for these biases to be present.

Even with the best possible analytical rigor, epidemiologic studies may have unavoidable limitations. A cross-sectional study may not be able to determine whether adverse environmental conditions led to increased health complaints or whether the health problems led to an increased perception of Physical discomfort. Interpretation of the directionality of the exposure-health effect association is based on a reconstruction of past events using current sources of information. Sometimes the investigators may not be able to make this determination.

The major limitation to epidemiologic studies of office environments is that in most buildings there are too few persons to reliably determine the association between specific exposures and health outcomes. Epidemiological analysis is based on the statistical comparison of disease rates in the study and comparison groups, controlling for other possible etiologic agents and biases. This analysis requires having a substantial number of subjects in the study. Furthermore, the number of subjects needed for the analysis increases as the number of possible etiologic agents increases. Thus, a very large number of subjects is needed to determine the cause of an office environmental problem with non-specific health complaints possibly being due to a variety of chemical, physical, and social agents. Rarely is this number of subjects present in one office building. This lack of sufficient numbers has hampered the ability of epidemiology to contribute more fully to the solution of these problems.

While epidemiologic techniques are particularly suited to the analysis of complex multifactorial problems, the number of subjects in the study must be larger than usually is available in office environmental problems. A possible solution to this dilemma would be for investigators to use standardized evaluation instruments (e.g., environmental measurements and questionnaires) and then to pool their data together to fully analyze the problem. I believe this approach should be considered by public health investigators involved in evaluating these office environment problems.

Conclusion

To date the fundamental etiology of office environmental problems has eluded detection. Extensive environmental monitoring has revealed the presence of myriad chemicals, but almost uniformly at extremely low air concentrations. Building inhabitants have associated their non-specific health complaints with perceptions of inadequate ventilation; however, ventilation rates rarely have been measured and are unlikely to be the sole cause of the problem. Social factors generally have been ignored and should be included in the evaluation of these problems.

Epidemiology can play a role in evaluating office environmental problems by complementing the activities of the industrial hygienist and ventilation engineer. The basic strategy of the epidemiological evaluation is to define the pattern of disease occurrence by time, place, and person. Recognition of disease patterns should generate hypotheses as to the etiology of the problem which then can be tested. During the evaluation, the epidemiologist should assess physical, organizational, and social factors not within the purview of industrial hygiene methods.

Epidemiological analysis should be used to identify causal associations between environmental exposures and disease rates. Analytical techniques are available to assess simultaneously the relative effects of multiple exposures and to control for potential biases. Unfortunately, these techniques require the participation of a larger number of study subjects than usually is available in a single office building. The full value of epidemiology in evaluating office environments will be realized only after standardized evaluation instruments have been developed so that the results of many investigations can be pooled together.

References

1. Kreiss, K. and M.J. Hodgson: Building-Associated Epidemics. *Indoor Air Quality*, pp. 87-106. P.J. Walsh, C.S. Dudney and E.D. Copenhaver, Eds. CRC Press, Boca Raton, FL (1984).
2. Fannick, N. and D. Baker: *National Broadcasting Company, New York, New York*. Health Hazard Evaluation Report No. HETA 81-417. National Institute for Occupational Safety and Health, Cincinnati, OH (1982).
3. Cohen, B.G.F.: *Human Aspects in Office Automation*. Elsevier Science Publishers, B.V., Amsterdam (1984).
4. Kasl, S.: Epidemiological Contributions to the Study of Work Stress. *Stress at Work*, pp. 3-50. C. Cooper and R. Payne, Eds. John Wiley & Sons, New York (1979).
5. Fannick, N. and D. Baker: *Maimonides Hospital, Brooklyn, New York*. Health Evaluation Report No. HETA 81-181-895. National Institute for Occupational Safety and Health, Cincinnati, OH (1981).
6. Baker, D. and N. Fannick: *United Nations, New York, New York*. Health Evaluation Report No. HETA 81-103-964. National Institute for Occupational Safety and Health, Cincinnati, OH (1981).

Application of health standards and guidelines

VERNON L. CARTER, JR., D.V.M.
College of Veterinary Medicine, The Ohio State University; Chairman, Chemical Substances TLV Committee

I feel that it is safe to say that almost all, if not all, occupational health guidelines and standards developed up to the present time have been generated for the industrial worker. This symposium is certainly timely in focusing our attention to the office environment so that those of us involved in establishing guidelines and standards may learn more about the problems and assist in offering a solution.

There are many sources of health standards and guidelines which may be used as a starting point. Federal regulatory and health agencies such as the Occupational Safety and Health Administration, the Environmental Protection Agency, the National Institute for Occupational Safety and Health and the National Institute of Environmental Health Sciences promulgate standards and publish recommendations which must be considered. The American Industrial Hygiene Association is very active in this arena with its Workplace Environment Exposure Limits. For many years the American National Standards Institute has also introduced health guidelines. Some foreign countries have developed standards; the most visible being the Federal Republic of Germany Commission for the Investigation of Health Hazards of Chemical Compounds in the Work Area. Two committees of the American Conference of Governmental Industrial Hygienists publish yearly recommendations on Threshold Limit Values (TLV®) of both Chemical and Physical Agents in the workplace. Finally, a number of industries have established company standards for a large number of compounds which have not been addressed by any of the afore mentioned agencies or groups.

In this presentation my remarks will be directed toward the use of TLVs in the office environment, although they will probably be applicable for all the occupational standards, recommendations and guidelines previously mentioned.

A review of the preface of the TLV booklet[1] would indicate that the TLVs are uniquely suitable for use in the office environment, since the eight-hour workday and the forty-hour workweek are standard in most office situations. The office environment does not pose the problem presently facing many of us as to how we develop standards and guidelines for non-traditional workshifts which are becoming more and more commonplace. There are, however, some potential problems that must be addressed in the use of TLVs in the office environment.

First, the TLVs assume a healthy worker. The preface to the TLV booklet[1] acknowledges that a small percentage of workers may be affected by aggravation of a pre-existing condition or by development of an occupational illness. Pre-employment and periodic physical examinations eliminate workers with health problems from workplaces which would be excessively hazardous during certain exposure situations. As an example, one would not place a worker with hepatic dysfunction into an environment of daily exposure to halocarbons. As a rule, office workers do not undergo the rigorous occupational medical examinations of industrial workers. Many office workers have existing medical problems, some of which they are unaware but many of which they are cognizant. Some employees may actually seek office positions because of pre-existing medical conditions which would preclude their working in industrial conditions.

Although the industrial worker, as with us all, suffers from some psychological stress, most health standards and guidelines address primarily the problem of physical stress. Although there certainly are exceptions, the usual industrial workday is organized and fairly routine. The industrial worker generally knows what he/she faces daily and what is expected of them. However, normal workloads and routines are the exception not the rule in many office environments. The sudden request for data, reports, briefing materials, etc. with ultrashort deadlines is a constant source of job related stress from middle managers to secretaries. The acute response to chemical agents in the psychologically stressed individual is certainly not well defined. In the search for etiology of an office health problem, care must be taken not to relate it to a chemical exposure only to discover subsequently that the real problem is the job related stressful office environment.

For particulate matter with a long history of little adverse effect on the lungs and which do not produce significant ongoing disease or toxic effect when exposures are kept under reasonable control, the TLV Committee has established a limit of 10 mg per cubic meter. Although this nuisance dust limit has been found to be acceptable in the industrial workplace, I doubt whether this value would be accepted in the office environment. Smoking is not allowed in most industrial situations for reasons other than health related effects. Tobacco smoke is offensive and irritating to many individuals. Smoking among a group of welders or coke oven workers on break would probably go largely unnoticed. This is not true in todays modern office environments. If such steps as arranging office smokers close to air exhaust ducts are not successful or are not feasible, than a ban on office smoking is the only practical solution. A "threshold limit" for passive tobacco smoke is not workable due to the extremely wide variation in the individual response.

Finally, one other major difference comes to mind between office and industrial workplace environments as we know them today. Most modern industrial environments present well defined contaminants. The chemicals used in the industrial situation and the products of industrial processes are known. The atmosphere can be sampled and analyses conducted for specific compounds. Previous presentations today have focused on the large number of sources of chemical agents in the office workplace. The problem of endeavoring to analyze qualitatively and quantitatively air samples which may contain many compounds is not appreciated among many health care professionals. In addition, the possibility of searching for one or more unknowns can lead to a process which can be both expensive and unrewarding.

In conclusion, there are two statements in the preface of the TLV booklet[1] which are important when applying health standards and guidelines in the industrial environment but become even more important when addressing the office environment. First *"the TLV-TWA should be used as guides in the control of health hazards and should not be used as fine lines between safe and dangerous concentrations."* The regulatory compliance arena has led many professionals to forget that the presence of a compound with a 1.0 ppm TWA may evoke a response at 0.9 ppm or may not have an effect at 1.1 ppm. Finally, *"these limits are intended for use in the practice of industrial hygiene and should be interpreted and applied only by a person trained in the discipline."* For those who are addressing the use of health standards in the office environment, it is going to require a patient and thorough examination of each situation and all the educational and practical experience faculties you can bring to bear. In most cases there will be no easy answers.

Reference

1. *Threshold Limit Values for Chemical Substances and Physical Agents in the Work Environment with Intended Changes for 1983-84.* American Conference of Governmental Industrial Hygienists, Cincinnati, OH (1983).

OFFICE BUILDING VENTILATION: DESIGN

Ventilation concepts for office buildings

PRESTON E. McNALL, Jr. and ANDREW K. PERSILY
Center for Building Technology, National Bureau of Standards

Introduction

The purpose of this paper is to describe the various commonly used ventilation systems for office buildings, and to discuss their general characteristics, advantages and disadvantages in providing the proper quality and quantity of ventilation to the occupied spaces. The emphasis will be on ventilation provided for the purposes of controlling the quality of the indoor air for the health and comfort of the occupants of the building. As shown, ventilation serves several other purposes in controlling indoor environments.

Ventilation is often described in terms of air changes per hour (ACH), even though ventilation is usually most important to people and is usually expressed for this purpose as a volume rate per person (cubic feet per minute per person, cfm/person, or liters per second per person, Ls^{-1} per person).

One ACH is defined as the volumetric flow of air provided to any space which would completely replace the volume of that space in one hour. ACH must then be related to occupant needs by determining the population density in the space, ft^2/person (m^2/person). Usually a 10 ft (3 m) ceiling height is employed. Office spaces vary in their density of population from about 75 ft^2/person (7.5 m^2/person) to 150 ft^2/person (15 m^2/person). Therefore, 1 ACH of outdoor air would provide a range of 12.5 cfm/person (6.2 Ls^{-1}/person) to 25 cfm/person (12.5 Ls^{-1}/person). Engineers sometimes specify ventilation as cubic feet per minute per square foot of floor area, cfm/ft^2 ($Ls^{-1}m^{-2}$) which can similarly be expressed in terms of cfm/person (Ls^{-1}/person) where the population density is known.

Purposes of ventilation

In office buildings ventilation serves several purposes. They are listed and discussed here.

1. To provide a healthy and comfortable air quality environment for the building occupants which will not be damaging to the building structure, furnishings, or contents. It is generally assumed that if people's needs are met, the building needs will be met as well. Exceptions would be archival storage and other special purpose areas whose contents require special consideration. These will not be covered here. The ventilation system provides air for comfort, which can be simply defined as air which is not subjectively objectionable to the occupants. Ventilation air is also provided for health reasons. Deleterious health effects can be either acute, or chronic. Since both health and comfort are ventilation concerns, minimum ventilation rates specified in building codes may be indicated for comfort or for health or both.

2. To provide for temperature control. Ventilation air is usually heated or cooled, and introduced into the space for the purpose of temperature control. Internal circulation is also needed to provide mixing for adequate temperature uniformity within any space. Practical limitations on temperature levels used in the heating, ventilating, and air conditioning (HVAC) systems usually require that the air flow quantities for temperature control be four to eight ACH. This is almost always the largest amount of ventilation required, and therefore becomes the design quantity. Ventilation for air quality rarely exceeds 1 ACH. Energy conservation considerations usually require that as much air as possible be recirculated, but recirculated air does not remove significant amounts of pollutants without special air treatment.

3. To control humidity. Moisture may be added to ventilation air using any one of a variety of humidifiers, and moisture may be removed by condensation on cooling coils, by chilled water sprays or chemical desiccants. Relative humidity levels are usually maintained below 60%, to protect the building structure from condensation, rot, etc., and to limit deterioration and mold growth on building contents. People can be thermally comfortable at relative humidities well above 60%.[1] In cold weather, relative humidities without humidification can be below

10%, and humidification is desirable to minimize discomfort, static electricity, etc. Condensation on cold windows and walls often prohibits the maintenance of relative humidities above 20% in cold weather.[1] Therefore, the practical range of relative humidities within buildings is 20-60%.

4. To remove excess heat, humidity, and contamination from the space. This is usually done by exhausting a portion of the air returned from the space before the balance is heated or cooled and returned to the space. In some cases special local exhaust systems are used. Office buildings are not usually equipped with special systems to serve zones of high internal heat or contamination rates with some exceptions such as lavatories and computer rooms. Computer rooms almost always have separate ventilating and air conditioning systems, and lavoratories are equipped with separate exhaust systems.

5. To provide small pressure differences between zones or between indoors and outdoors to control air exchanges, such as in entrance lobbies.

6. To provide pressure differences to confine or exhaust smoke, heat, and toxic fumes for fire safety. The HVAC system is being used in an increasing number of cases for fire safety. Special dampers and fans are programmed to respond to smoke and fire sensors in various locations so that when problems are detected, the ventilation system will not spread the products of combustion, but confine and exhaust the smoke and fumes according to strategies which provide "safe havens" and/or safe exit passages for the occupants.

7. To provide enhanced air motion in hot spaces for greater thermal comfort. This is not usually done in offices.

General ventilation concepts

Early office buildings had no mechanical ventilation systems. Air conditioning was not provided and heat was distributed using steam or hot water in radiators, etc. Ventilation was provided by open windows and cracks around windows and doors, and through vertical ventilation shafts. The assumption was made, and is usually made today, that outdoor air is the "contaminant free" source of ventilation air. Outdoor air used to be termed "fresh air" by heating, ventilating, and air conditioning engineers. Recently, it has been realized that outdoor air is not always "fresh," and often must be filtered before it is suitable for use as ventilation air. The term "outdoor air" is now used.

As central air conditioning systems became popular, mechanical systems were needed, and the concept in recent years has been for the fan system to distribute the air. Outdoor air is introduced from one or several locations and distributed through duct systems to the spaces. Return air is removed through separate duct systems, and some is exhausted through exhaust outlets.

The ventilation system is designed to deliver this outdoor air to the spaces in order to meet the appropriate ventilation requirements. The effect of air leakage through the building enclosure is ignored. This leakage adds a variable and unknown quality of outdoor air to the ventilation in the spaces provided by the fan system, increasing the ventilation rate and providing a "factor of safety."

Since air leakage (infiltration) is uncontrolled and has a significant energy cost when heating and cooling is required, recent building design and construction practices attempt to reduce infiltration as much as possible. However, no building enclosure can be made air tight. Infiltration depends upon the enclosure construction, and on the wind conditions, which cause pressure differences on the building surfaces. In addition, indoor-outdoor temperature differences cause a "stack-effect" due to air bouyancy differences between indoors and outdoors. This is particularly noticeable in tall buildings. Therefore, infiltration varies considerably for any building from time to time, depending upon the weather.

Recent studies[2,3] on several government office buildings, most of which were constructed within the last five years, show air change rates for periods of maximum heating and cooling loads from 0.2 ACH to 0.7 ACH. Values at any one time may be somewhat larger or may approach zero, depending on the weather and internal conditions.

General design concepts

Commercial buildings are usually large. Therefore, the building heating, ventilation and air conditioning systems (HVAC) are usually designed to serve the building core differently from the perimeter. The core, due to internal loads of people, lights, etc., has only a cooling load during occupied hours. Heating is provided only during unoccupied hours and for morning warm-up periods. The core or interior of small buildings may be treated as a single zone for temperature control purposes, but in large buildings several separate control zones are provided, allowing for various occupancy levels and uses in the different zones.

The perimeter volume usually extends from 10-15 feet (3-5 m) from the exterior walls[1] and must be provided with additional heating and cooling to offset the heating or cooling transmission loads through the building enclosure as weather conditions change. Solar and wind effects are different on the various faces of the perimeter, so these areas are usually divided for control purposes into four or more zones according to the orientation of the building. These zones may have hydronic heat in the form of finned tube radiation which provides no ventilation. They may have separate zones from a central air supply or they may be provided separate air systems altogether. Small office buildings may have all perimeter systems with no core system. Figure 1 shows a floor plan of a typical building with some typical temperature control zones.

Specific ventilation systems

The following are basic ventilation systems which are commonly used in new and existing buildings. There are many variations of these systems and many different control strategies in use, particularly those which are more energy efficient. Each system's ability to provide adequate indoor air quality is discussed. The basic reference material for this information is the American Society of Heating, Refrigerating and Air-Conditioning Engineers Handbook Series.[1]

Central single zone constant volume systems

Figure 2 shows a schematic of this commonly used system. Heating and cooling heat exchangers provide for energy transfer. Heating may be provided by a steam or water boiler, electric heat, heat pump, etc. The cooling may be provided by a vapor compression air conditioner, absorption air conditioner, or water chiller. Energy conservation practices have resulted in a number of heat reclaim methods which can assist in producing economical heating or cooling. Subsequent systems discussed may also employ one or more similar energy sources.

The supply fan distributes the air to the space through a duct system and air is returned through a separate duct system using a return air fan. In small systems, there may be no return air fan. The return air is partly exhausted to offset the amount of outdoor air introduced for air quality purposes and the balance is recirculated. A zone thermostat controls the energy supply to the heating or cooling coil as required for temperature control. There are always filters of some type preceding the coils, but these are usually low efficiency filters for the purpose of protecting the coils, fan and the rest of the system from fouling rather than for the purpose of improving air quality. The dampers shown are operated at positions which are designed to admit the required amount of outdoor air to satisfy the air quality needs of the building as specified in the applicable building codes. The systems are usually designed so that the exhaust and outdoor air dampers may be closed during unoccupied periods to save energy.

This system and other systems described later may also employ an economizer cycle. This control strategy senses when the outdoor air has a low enough heat content to permit it to be used instead of mechanical cooling to save energy. Usually when the outdoor air is between about 55°F (13°C) and 70°F (21°C) it can be used to cool, or assist in cooling, the spaces. Then the by-pass

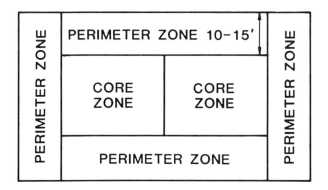

Figure 1 — Typical office floor temperature control zones.

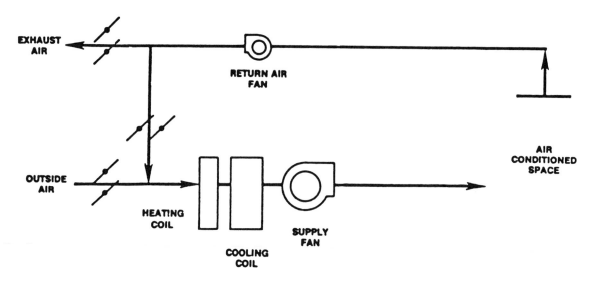

Figure 2 — Single zone constant volume system.

dampers are closed and the outdoor air and exhaust air dampers are opened to take as much as 100% outdoor air through the system. A more sophisticated control, called an enthalpy controller, can be used to also sense the moisture content of the indoor and outdoor air to provide for additional energy saving. During such periods, more outdoor air than the minimum is supplied to the space.

Central systems may be very large, supplying several floors of a large building when properly zoned. The return air plenum upstream from the coils may be 10 ft × 20 ft (3 m × 6 m) in cross section or even larger. For single story buildings, the entire system, up to 1 million Btu/hour capacity (293 kilowatts) or so of heating and cooling may be factory assembled in a roof-top unit, which is complete with its own heating and cooling sources. At the building site, only the duct work and the necessary zoning controls need to be added.

A humidifier may also be included in the system. Dehumidifying is accomplished by condensing moisture from the air on the cooling coil.

There are practical problems associated with this system's ability to provide outdoor air for air quality purposes. First, the air dampers are not precision control devices and it is difficult to determine if the intended minimum quantity of outside air is actually being introduced into the building. Second, the duct work leaks and, therefore, flow rates may not be the design values in some spaces. Leakage for well constructed and sealed duct work is as low as 1-3%,[1] but it may run to 20-30%[1] in poorly constructed ducts. Third, the supply outlets in the various spaces may not deliver the design quantity of air to each space at all times due to the action of the control system at various loads. Fourth, filter loading, system fouling, and wear and tear over time will affect the air flows.

Another recently noted problem in all ventilating systems is the degree of air mixing in the space. Space mixing is controlled by the design and location of inlets and outlets in the space as well as the quantity of air flow to the space. Over the years devices and installation practices have been developed to provide acceptable temperature distribution in spaces, but very little research has been done which considers the contaminant distribution in the spaces. Economics and the flexible use of office space have usually dictated that both the air inlets and outlets be located in the space ceiling. How much by-passing of the ventilation air occurs in various space distribution systems and how much this increases the contaminant levels above the assumed values in the occupied zone is a fertile area of research.

System balancing is difficult, and the balance may change with various operating conditions. Space modifications are often made to suit different tenants and these modifications may also alter the system balance.

Ventilating systems themselves may cause additional problems. Cooling coils condense moisture from the air and require some form of drain. Drain pans may be poorly installed or maintained and standing water may be present. Humdifiers may introduce water sprays or boil water from a tray in the ductwork. Standing water may also be present in or near these. Duct work may have construction debris, even remnants of workers lunches, left in it during construction, and other debris can accumulate over time. Such debris and moisture may provide areas in the system where microorganisms can deposit and multiply, to be later entrained in the ventilation air.

Recirculation of a portion of the exhaust air into the outdoor air intake may also cause problems, as well as recirculation from lavoratory exhausts, kitchen exhausts, etc. Other contaminants from local vehicle traffic or nearby building exhausts may also be introduced in to the outdoor air intake at various times.

These general charactistics of the single zone systems usually also apply to those described below.

Terminal reheat constant volume systems

Figure 3 shows a schematic of a terminal reheat constant volume system. It is similar to Figure 2 except the supply duct work is arranged to feed several separate space zones. Each zone has a reheat coil controlled by a zone thermostat. These systems are notorious energy users since the central systems supply cool air, usually employing energy, to all zones and then much of the air is reheated, often at a double cost of energy. Many of these systems exist, since they provide excellent temperature control and were designed when energy costs were not as important as now. Their ventilation problems are similar to those of single zone systems.

Dual duct constant volume systems

Figure 4 shows a schematic of a dual duct constant volume system. Here separate hot and cold ducts (called hot and cold decks) are run to the control spaces and each zone thermostat controls the portion of hot and cold air to be mixed to satisfy the temperature needs of the zone. In addition to the ventilation problems referred to earlier the flow rates in the hot and cold ducts fluctuate as the loads change, adding some

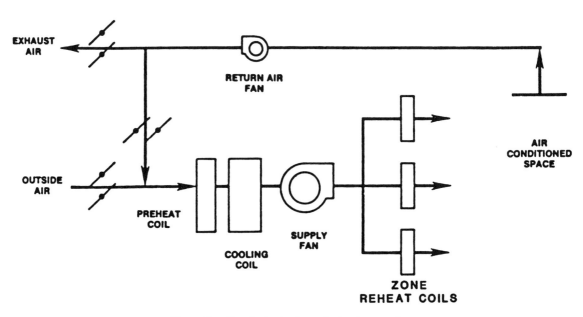

Figure 3 — Terminal reheat constant volume system.

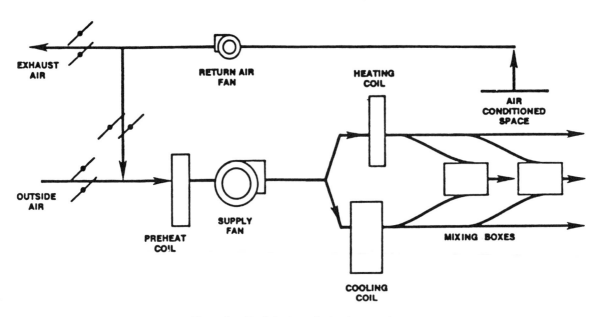

Figure 4 — Dual duct constant volume system.

uncertainties concerning the ventilation in each space. As with the terminal reheat systems, their energy use is high, and many retrofit methods to save energy have been devised.

Variable air volume systems

Figure 5 shows a schematic of a simple variable air volume (VAV) system. Instead of varying the temperature of the air to the space, the air volume is reduced at part load. These systems are common now since they have great potential for saving energy. Each variable volume box has a minimum setting which can be adjusted for the purpose of establishing the minimum requirement for ventilation for air quality in each zone. However, since the outside air is usually only a portion of the total supply air, the difficulty of ensuring that this required minimum outdoor air rate is supplied to each zone at all times is apparent. Such systems also require methods of modulating the flow through the supply and return ducts to provide proper pressure control in the spaces. This is usually done by special fan controls.

Double duct variable air volume systems

Figure 6 shows a double duct variable air volume system schematically. The availability of both heated and cooled air to each zone gives the opportunity for better temperature control.

Variable air volume reheat systems

Figure 7 shows another variation of the popular VAV system with terminal reheat for better and more flexible temperature control.

Induction systems

Figure 8 shows a typical induction system. In these systems, only the minimum amount of ventilation air is provided by the central system. This saves space and materials, employing smaller ducts, and can also save fan power. Additional means of heating and/or cooling must be provided in each terminal unit and additional air is induced from the space as shown to provide better mixing in the space and the increased total quantity of air necessary to satisfy the heating and cooling loads.

Perimeter systems

As previously noted, the perimeter system may be zoned as a part of a central system or it may be separate. In some cases the perimeter will be heated by finned tube radiation at the floor level. This system does not provide any ventilation air. The ventilation air is then supplied by any of the ventilation systems covered previously or subsequently.

Another perimeter system is the fan coil system shown in Figure 9. These are very common in

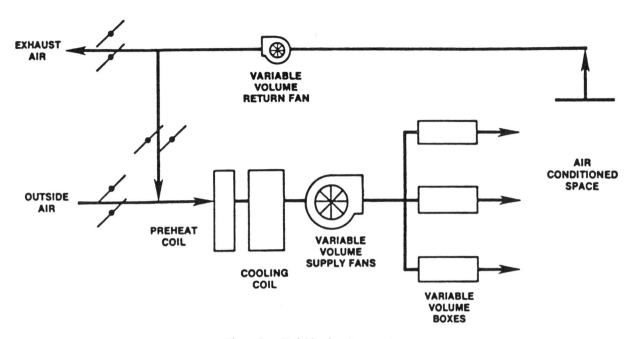

Figure 5 — Variable air volume system.

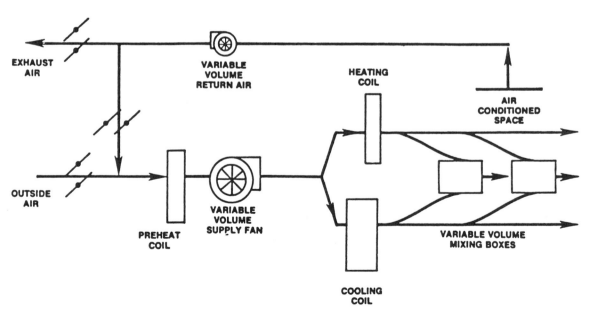

Figure 6 — Double duct variable air volume system.

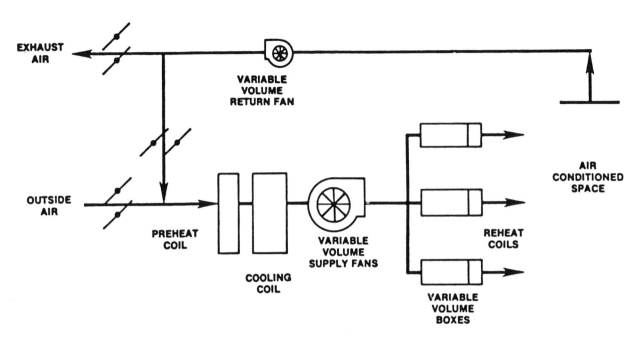

Figure 7 — Variable air volume reheat system.

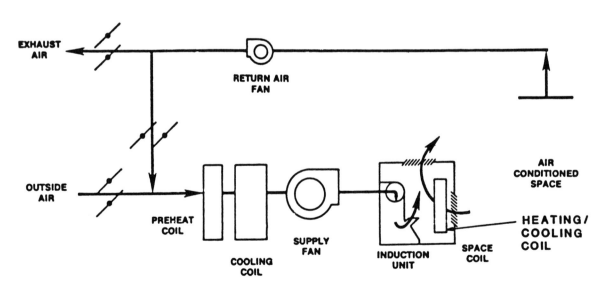

Figure 8 — Induction system.

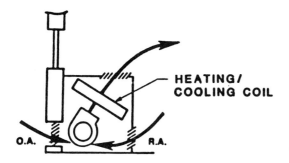

Figure 9 — Fan coil unit.

motels, hotels, and small office buildings. These systems may provide a small amount of ventilation from outdoors as shown, or there may be no outdoor inlet. They may provide for water heating and cooling from a central source, or they may be self-contained, employing a small air conditioner or heat pump and have supplemental electric heat.

The perimeter may also be served by an air system as shown in Figure 10. The one shown may be variable volume or constant volume, and provides no outdoor air. The return duct will collect some additional air from the interior zones and hence, some is returned to the central system.

Summary

- There are many systems used to ventilate office buildings, and there are many variations of these used to save energy and to provide for more sophisticated temperature control.

- Energy concerns have resulted in many new designs and retrofits which reduce ventilation to lower levels than when energy was less expensive.

- In many systems it is difficult, if not impossible, to ensure that any particular zone receives the specified amount of outdoor air for air quality purposes at all times.

- In recent years, energy concerns have led to tighter building enclosures which reduce uncontrolled infiltration. Energy concerns also tempt designers to use the lowest possible ventilation rates. Both of these trends exacerbate air quality problems in the buildings.

- Inadequate mixing in the space probably often occurs. Such inadequate mixing increases pollutant levels which cause problems.

- Poor construction and maintenance of duct systems may provide environmental niches for the growth of undersirable microorganisms which can be entrained in the duct air and enter the building spaces.

Recommendations

This paper suggests the following recommendations:

1. Better ventilation standards are needed.
2. Research is needed to provide better guidelines for component and system de-

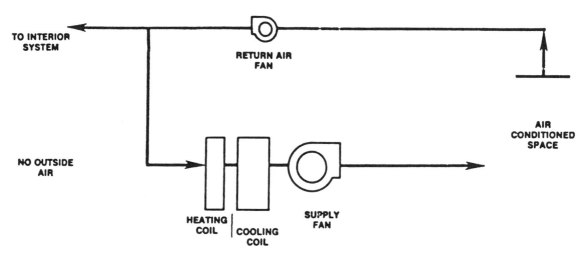

Figure 10 — Perimeter air systems.

sign so that better ventilation systems can be designed.

3. Research is needed to provide better guidelines for operation and maintenance of ventilating systems.

4. Research is also needed to provide better methods for the field evaluation of ventilating systems to ensure that they perform as intended.

References

1. *American Society of Heating, Refrigerating, and Air-Conditioning Engineers Handbook*, Volumes on *Systems* (1980), *Fundamentals* (1981), *Applications* (1982), and *Equipment* (1983). Available from ASHRAE, 1791 Tullie Circle, NE, Atlanta, GA 30329.

2. Grot, R.A. and A.K. Persily: Air Infiltration and Air Tightness Tests in Eight U.S. Office Buildings. *Air Infiltration Reduction in Existing Buildings*, 4th AIC Conference, Elam, Switzerland (September 26-28, 1983).

3. Grot, R.A., Y-M.L. Chang, A.K. Persily and J.B. Fang: *Interim Report on NBS Thermal Integrity Diagnostic Tests on Eight GSA Federal Office Buildings*. NBSIR 83-2768. Available from NTIS, Springfield, VA 22161 (September 1983).

Ventilation for acceptable indoor air quality: ASHRAE Standard 62-1981

JOHN E. JANSSEN*
Honeywell, Inc., St. Paul, MN

Introduction

It has been suggested that early man survived among larger and stronger animals partly because he had a disagreeable body odor. When he brought his campfire into his cave, he compounded his air pollution problem. Indoor air pollution is not a new problem.

There are three options for controlling air pollutants or contaminants. The offending pollutant may be removed or eliminated at the source. Venting of the products of combustion from a furnace, use of exhaust fans in kitchens and bathrooms, and substitution of materials which do not emit undesirable pollutants are examples of source control.

If the pollutant cannot be kept out of the indoor atmosphere, in some cases it can be removed by filtering. Electrostatic and media filters are widely used for removing particulate matter from the indoor air. Electrostatic filters are particularly effective for removing small particles in the size range of pollens. There is some evidence that removal of particulates also removes radio active decay products of radon gas. These decay products are radio active isotopes of certain metal atoms. They tend to be attracted to small particulates by electrostatic forces and thus are removed from the air when the particulates are removed. One shortcoming of the filtration option is the general requirement for regular maintainance.

The third, and most widely used control strategy, is dilution of polluted indoor air with less polluted outdoor air. The ASHRAE Ventilation Standard addresses the requirements for proper and effective use of this method. However, source control and filtration each have certain advantages which can and should be exploited in any building. This ventilation standard recognizes and recommends their use where appropriate.

In reviewing the current status of ventilation technology and the ASHRAE ventilation standard in particular, it is helpful to consider the past history. The effect of the 1973 oil embargo and escalating fuel prices have increased the demands on ventilation systems. A careful comparison of the 1973[1] and 1981[2] versions of ASHRAE Standard 62 shows more similarity than difference. The current effort to rectify opposition to the '81 version is partly directed at making the standard more easily understood. Specific attention is directed at the problem of controlling tobacco smoke, problems of ventilation efficiency and at the use of the alternate "Air Quality Procedure."

History

Woods,[3] in an interesting historical review of ventilation technology, found the earliest effort to define ventilation requirements dated from 1836, Tredgold found that 2 L/s (4 cfm) per person of outdoor air was needed to dilute the CO_2 exhaled by occupants in a space. The recommended ventilation rate increased as central heating and mechanical ventilation systems become more popular. By 1905 the recommended outdoor air flow rate was 30 cfm per person. This high dilution rate was needed to control human body odors. Personal hygiene improved during the early part of the 20th century and by the 1930s, Yaglou and his associates established 5 L/s (10 cfm) per person as the basic recommended outdoor air flow rate for ventilation of buildings. This rate was still based on odor control. The first ventilation standard, in the United States was published in 1946. The "American Standards Building Requirements for Light and Ventilation — A53.1," was published by the American Standards Association.[4] This standard remained in effect until it was updated by ASHRAE in 1973. ASHRAE Standard 62-73, "Standards for Natural and Mechanical Ventilation," was also adopted by the American National Standards

---

* Mr. Janssen is also Chairman of ASHRAE Committee SPC 62-1981R that is reviewing Standard 62-1981, *Ventilation for Acceptable Indoor Air Quality.*

Institute (ANSI, formerly ASA) in 1977 and was given the designation ANSI Standard B194.1. Standard 62-73, for the first time, gave a quantative description of "acceptable outdoor air" and specified conditions for use of recirculated air.

Table 3 of Standard 62-1973 presents both recommended and minimum outdoor air flow rates for a variety of applications. The recommended flow rates considered odor and other comfort related factors. The minimum flow rates were based on the amount of outdoor air needed to dilute occupant generated carbon dioxide to an acceptable concentration of 0.25% under steady state conditions.

Energy concern

Unfortunately the ink on the '73 standard was barely dry when the oil embargo suddenly cast new emphasis on energy. Subsequently when ASHRAE Standard 90-75, "Energy Conservation in New Building Design,"[5] was published in 1975, it recommended use of the minimun flow rates specified in Standard 62-73. An objective, when Standard 62 came up for review in 1978 was to rectify any differences between the objectives and specifications of the two standards.

It was known that the energy required to heat and cool outdoor air for ventilation of buildings amounted to between 20 and 50% of the energy used in many buildings. Infiltration losses in homes amounts to 20-40% of the energy needed to heat or cool a home. The incentive to conserve energy following the oil embargo led building operators and builders to modify ventilation systems without regard for the consequences. At the same time many new materials were being introduced into the building industry. Thus, recommended ventilation practices were ignored while at the same time the contaminant load was being increased. The result has been the "Sick Building Syndrome."

Standard 62-73 vs. Standard 62-1981

When Standard 62-1981 was submitted for public review in the spring of 1979, approximately 30 people raised about 180 questions. These were all addressed by the committee, changes were made, and all but 9 were resolved. The unresolved issues were considered to be inapplicable, trivial or incorrect and were rejected. The ASHRAE Standards Committee accepted this resolution of issues and voted to recommend approval to the ASHRAE Board of Directors. The ASHRAE Board also approved the standard, but before it was published, the manufactured housing industry suddenly came forward with a complaint that the formaldehyde limit recommended in the standard was indefensible and would ruin the industry. This group had not raised the issue during the public review. ASHRAE, in an effort to be fair, instituted a new appeals procedure. A hearing was held in November 1980 at which time the Formaldehyde Institute presented its arguments and the ASHRAE Standard Project Committee presented its rebuttal. The result was that the Formaldehyde Institute appeal was denied and ASHRAE published the standard in 1981. The American National Standards Institute, however, declined to approve the standard because of the controversy.

Acceptable outdoor air

Table I presents a comparison of the recommended limits for certain contaminants in outdoor air that is to be used for ventilation purposes. The 1981 figures are the National Ambient Air Quality Standards. There are some small differences between the 1973 and 1981 publications, but the differences are small. The particulate levels were actually relaxed in the 1981 version. Both standards state that outdoor air used for ventilation should meet the conditions listed in Table I,[1,2] or be treated so that it does. This requirement has not been questioned in either standard.

Both standards then define governmental authorities and methods for certifying that the local outdoor air does meet these standards for acceptability of the outdoor air. Both standards further recognize that there may be other contaminants with which to be concerned. Standard 62-73 does not provide any guidance as to what other contaminants should be considered, but instead says,

> "3.3 AIR shall be considered unacceptable for ventilation use in accordance with this standard if it contains any contaminant in a concentration greater than one-tenth the Threshold Limit Value (TLV) currently accepted by

TABLE I
Acceptable Ventilation Air

Contaminant	Annual Average		Short-Term Level			
	1973 mg/m^3	1981 mg/m^3	1973 mg/m^3	time	1981 mg/m^3	time
Carbon monoxide	20,000	—	30,000	8 hr	40,000	1 hr
Lead	—	1.5 for 3 mo.	—	—	—	—
Nitrogen dioxide	200	100	—	—	—	—
Oxidants (ozone)	100	—	500	1 hr	235	1 hr
Particulates	60	75	150	24 hr	260	24 hr
Sulfur dioxide	80	80	400	24 hr	365	24 hr
Hydrocarbons	1800	—	4000	3 hr	—	—

the American Conference of Governmental Industrial Hygienists."

Standard 62-73 goes on to specify procedures for sampling and analysis. Standard 62-1981, on the other hand, presents a table (Table 2 of the standard) of "Additional Ambient Air Quality Guidelines." This table, which covers common contaminants for which no EPA ambient air quality standard exist, presents data selected from current practices in various states, provinces, and other countries. The designer is also cautioned to consider any other contaminants which he may suspect exist.

The additional air quality guidelines were put in Standard 62-1981 to aid the designer in unusual cases. These materials would not normally be found unless the location is near a chemical plant, an industrial complex or some potential source of industrial process materials. The response of engineers has been opposite from the intention, however. Since there was no listing of additional contaminants in Standard 62-73, they went largely ignored. Listing the potential hazards in a table in the 1981 standard suddenly made everyone believe that these all had to be measured. Measurement procedures were unfamiliar to most ventilation system designers and they felt threatened.

The present review committee is considering several options. One is to put the table of Additional Ambient Air Quality Guidelines in the appendix. This would remove it from the standard but would preserve the information. It is recognized, however, that a ventilation system designer must exercise caution to assure himself that the air used for ventilation of a building does not contribute to deterioration of the indoor air quality. The outdoor air must be treated to remove unacceptable contaminants if it is not inherently acceptable.

Minimum outdoor air requirement

The primary indoor air contaminant requiring dilution is carbon dioxide exhaled by the occupants of a space. Appendix D of Standard 62-1981 presents the rationale for the minimum outdoor air dilution rate based on the CO_2 level.

The oxygen used in the metabolism of food converts hydrogen to water and carbon to carbon dioxide. The hydrogen carbon ratio in foods varies somewhat. The respiratory quotient (RQ) is the volumetric ratio of carbon dioxide produced to oxygen consumed. It varies from 0.71 for a diet of 100% fat, to 0.80 for a diet of 100% protein, and 1.0 for a diet of 100% carbohydrate. A value of RQ = 0.83 has been found to apply to a normal mix of fat, protein and carbohydrate in the average diet.

The oxygen consumed also depends on activity level. A sedentary adult consumes about 0.0059 L/s (0.0125 cfm) of oxygen. Assuming about 25% of the O_2 inhaled is used by the body, and knowing that air is about 21% O_2, the air required by a sedentary adult for oxygen supply is:

$$[(0.0059)/(0.21)(0.25)] = 0.11 \text{ L/s } (0.24 \text{ cfm}) \quad (1)$$

The controlling physiological factor is the CO_2 concentration in the inhaled air rather than the oxygen depletion. Men can function at CO_2 levels as high as 1% in nuclear submarines. Studies in bomb shelters showed that 0.5% CO_2 was a feasible level for healthy individuals. On this basis a level of 0.25% was selected as acceptable for general building environments. There is approximately 0.03% CO_2 in the atmosphere, and the rate of CO_2 production for a sedentary adult is 0.005 L/s person (0.0105 cfm/person). A simple mass balance shows that (Equation E-1, Appendix E, Standard 62-1981):

$$\dot{V}_o C_o + \boxed{\dot{N}} \rightarrow \dot{V}_o C_i \qquad (2)$$

$$\dot{V}_o = \frac{\dot{N}}{C_i - C_o}$$

where:
- $\dot{V}_o$ = Air flow rate
- C_i = Indoor CO_2 concentration (0.25%)
- C_o = Out CO_2 concentration (0.03%)
- $\dot{N}$ = CO_2 generation rate (0.005 L/s)
- $\dot{V}$ = 0.005/(0.0025 − 0.0003) = 2.27 L/s (4.82 cfm)

Thus, it takes 20 times as much air to satisfy the CO_2 dilution requirement as it does to satisfy the O_2 depletion need.

Table II shows the outdoor air needed to maintain various steady state levels. The values have been rounded off and a converson factor of 1 cfm = 0.5 L/s has been used. This is consistent with Standard 62-1981.

Table II provides tha basis for the Ventilation Rate Procedure of Standard 62-1981. The basic rate of 2.5 L/s which yields a steady state CO_2 level of 0.25% was chosen. This is consistent with the minimum outdoor air flow rate specified in Standard 62-73. It is also consistent with the calculations of Tredgold[2] made in 1836.

The smoking issue

The two standards provide increased outdoor air for various applications where activity may be greater than sedentary or other contaminants may be expected. Table III presents a brief comparison of the two standards. Standard 62-73 presented both minimum and recommended outdoor air flow rates. Standard 62-1981 generally endorsed the minimum flow rates presented in the 1973 standard. Smoking was recognized as a special problem, however. Thus, the 1981 version defined the outdoor air requirements for smoking or non-smoking areas. This has been a source of misunderstanding and concern.

The minimum flow rates in the '73 standard are very similar to the non-smoking flow rates in the '81 version. The 1973 standard identified smoking specifically in auditoriums and specified twice the outdoor air when smoking was permitted. In other cases where smoking is common, e.g., bars, the minimum outdoor air flow rates specified in Standard 62-73 were only slightly less than the recommended rates. Thus, Standard 62-73 tacitly assumed that the recommended columns would be used where smoking was allowed.

In general offices spaces and bars, for example, the recommended column and the smoking column, as shown in Table III are similar. Standard 62-1981 did define some of the conditions differently, however. The outdoor air requirements for hotel living rooms and bedrooms are given in L/s (cfm) per room instead of per person. Since occupant density is quite variable but room sizes are relatively standard in these cases, the committee felt a flow rate per room was more logical.

A ventilation system designer usually determines the occupant density per unit area floor space and multiplies this by the flow requirement per occupant to get a flow rate per unit floor area. Outdoor air requirements for general office spaces and for bars are compared in Table IV.

The greater occupant density along with an assumed increase in smoking in bars greatly increases the outdoor air requirements. Table IV was calculated using the recommended column from Standard 62-73. Minimum flow rates would

TABLE II
Outdoor Air Requirements for Various Steady State Indoor CO_2 Levels
[sedentary adult activity assumed]

% CO_2 Limit	Outdoor Air Flow Rate	
	L/s	cfm
0.5	1.25	2.25
0.4	1.43	2.86
0.3	1.96	3.92
0.25	2.50	5.00
0.2	3.12	6.23
0.1	7.55	15.10

TABLE III
Outdoor Air Requirements

Application	Occupancy Occup. per 100 m² or per 1000 ft	Minimum L/S • Person (cfm)	Recommended L/S • Person (cfm)	Occupancy Occup. per 100 m² or 1000 ft²	Smoking L/S • Person (cfm)	Nonsmoking L/s • Person (cfm)
General office	10	7 (15)	7-12 (15-20)	7	10 (20)	2.5 (5)
Auditoriums	150	2.5 (NS) (5)	2.5-5 (NS) (5-10)	150	17.5 (35)	3.5 (7)
Hotel living rooms	20	5 (10)	7.5-10 (15-20)	20	25* (50)	12.5* (25)
Hotel bedrooms	5	3.5 (7)	5-7.5 (10-15)	5	15* (30)	7.5* (15)
Bars	150	15 (30)	20-25 (40-50)	100	25 (50)	5 (10)
Residential living areas	5	2.5 (5)	3.5-5 (7-10)		5* (10)	5* (10)
Industrial facilities (occupants only)	—	3.5-17.5 (7-35)	5-22.5 (10-45)	—	17.5 (35)	10 (20)

* Airflow per room instead of per occupant.

TABLE IV
Outdoor Air Required for Smoking

	L/s × m²	(cfm/ft²)
Standard 62-73		
Application:		
General office	1.2	(0.25)
Bars	37.5	(7.50)
Standard 62-1981		
Application:		
General office	0.7	(0.14)
Bars	25.0	(5.00)

have been 40% less. Thus, Standard 62-1981 gives outdoor air requirements for smoking which are generally between the minimum and recommended values in Standard 620-73. This varies by application.

The outdoor air flow rates for dilution of tobacco smoke presented in Standard 62-1981 are based on the control of particulates since they constitute the largest mass of pollutants in the smoke.[3] Equation 2 holds for particulates as well as gaseous contaminants.

The average generation rate for total suspended particulates is 31.9 mg per cigarette.[6] The National Ambient Air Quality specifies 0.260 mg/m³ as the concentration limit for total suspended particulates for 24-hour exposures.[2] If the outdoor concentration of particulates is assumed to be zero, the outdoor air flow rate needed is:

$$\dot{V}_o = 31.9/(0.260 - 0) = • 123 \text{ m}^3/\text{hr} • \text{cigarette} \quad (4)$$

If we assume 30% of the occupants in a room smoke 2 cigarettes per hour, the outdoor air needed is:

$$\dot{V}_o = (0.30)(2)(123)(100)/3600 = 20.5 \text{ L/s (41 cfm)} \quad (5)$$

The flow rates recommended in Table 3 of Standard 62-1981 were adjusted up and down from this basic rate. Smoking in areas such as bars tends to be greater than the assumed rate, but the duration of exposure is substantially less than 24 hours. There are always some particulates in outdoor air which would increase the amount of outdoor air needed. The National Ambient Air Quality Standard specifies 0.075 mg/m³ as the maximum particulate concentration for continuous (1 year) exposure. There is also a finite settling rate that removes some of the particulates. The recommended outdoor air flow rate considers these factors.

Appendix E of Standard 62-1981 also shows how filters can greatly reduce the amount of outdoor air needed. The total circulation rate must be increased, however. This may present a problem in variable ventilation systems.

Variable occupancy

Standard 62-1981 introduced additional features not contained in Standard 62-73. One of these was a provision for variable occupancy. It was recognized that many buildings are not occupied continuously. Schools, office buildings, etc. are not occupied at night. If the contaminants are generated primarily by the occupants, e.g., CO_2, moisture, odors, tobacco smoke, the building tends to be cleared of contaminants at night when it is unoccupied. It is feasible, therefore, to delay the start of the ventilation system in the morning until after the occupants have entered the building and contaminant levels have risen to the "control" point. Figure 1 presents a graph from Standard 62-1981 for determining the maximum permissible ventilation lag time. A permissible delay of one hour in the start up of the ventilation system is not uncommon.

If contaminants are generated by the space itself, however, start up of the ventilation system must lead the occupancy period. Outgassing of formaldehyde, organics and such that is independent of occupant activities will cause increased concentrations during unoccupied periods when the building ventilation system is off. The system must then be started in advance of occupancy in order to bring contaminant levels down to an acceptable range. Figure 2 (Figure 3 in Standard 62-1981) shows the lead time required. Lead times of the order of one hour are probably common.

Air quality option

Section 6.2 Indoor Air Quality Procedure of Standard 62-1981 is an alternative procedure. Standard 62-73 had no similar provision. This air quality procedure was included as an alternate to permit and encourage innovative, energy conserving solutions to the problem of building ventilation. The procedure, stated simply, says the designer may use any amount of outdoor air he

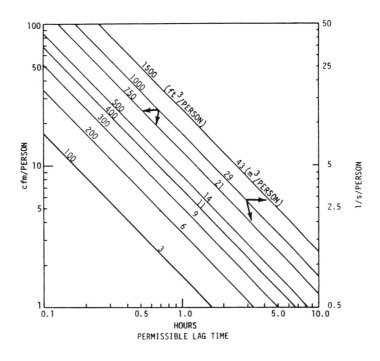

Figure 1 — Maximum permissible ventilation lag time.

wants to use if he can show that specified contaminants are kept below the specified limit. Table 4 of Standard 62-1981[2] specifies recommended concentration limits for only 5 contaminants. The standard also advises the designer to consider contaminants listed in Table 2 of the standard. Thus, there is a considerable burden on the engineer to use the Air Quality Procedure. The committee recognized this problem but felt the option was desireable.

Unfortunately the procedure has been misunderstood and misinterpreted. Standard 62-1981 recommends a formaldehyde limit of 120 $\mu g/m^3$ (0.1 ppm). This was based on evidence that shows that sensitive people begin to experience irritation of the eyes and mucous membranes at about this level. Formaldehyde is known to cause cancer in laboratory animals at some higher concentration, but the risk to humans is unknown. Standard 62 is concerned with ventilation for comfort as well as prevention of health risk. Thus, the recommended formaldehyde limit is a comfort criterion rather than a health risk criterion.

Standard 62-1981 plainly states that the Air Quality Procedure is an alternate procedure. If the ventilation rate procedure is followed, it is assumed (ipso facto) that the indoor air quality will be acceptable. The main problem arises with residential applications and in particular manufactured housing. Residential systems do not use mechanical ventilation, but depended instead on passive infiltration. Newly manufactured homes use many wood products bonded with urea formaldehyde resins. These emit formaldehyde vapors, especially newly manufactured materials. Fabric finishes draperies, carpeting, and particle board used in furniture add to the formaldehyde load. The manufactured housing industry feared that they would be blamed for the entire formaldehyde problem even though their construction materials were only part of the formaldehyde source.

Since building inspectors had no good way to measure infiltration, but they could measure formaldehyde concentration. This could become the controlling factor even though the standard did not intend it to be. This has been the core of the argument over formaldehyde.

Current status of standard 62-1981

An ASHRAE Standards Project Committee began a review of Standard 62-1981 in January of 1983. Several areas are of particular concern. One of these is the issue of odor. Work by Dr. William Cain and his associates at the John B. Pierce Foundation, Yale University, suggests that the minimum outdoor air flow rate of only 2.5 L/s person (5 cfm/person) will satisfy only about 50% of the visitors to a space from the standpoint of odor. Occupants rapidly become adapted and cease to notice odors. Visitors coming from a "clean" area, however, will notice odors on first entering a space. Cain estimates that it will be necessary to increase the outdoor air flow rate from 2.5 L/s person to 7.5 L/s in order to reach 80% acceptance by visitors. This raises a question of whether we should ventilate for visitor or adapted occupants. One approach being considered is to define two classes of spaces. Hotel lobbies, waiting rooms, retail stores, etc. where occupancy is transient may be ventilated for visitors. Offices, hotel bedroom, etc. where occupancy is generally more stable could be ventilated for adapted occupants. This approach is under consideration.

The smoking issue is also under study. Recent work by Repace[7] at EPA has produced a correlation between risk and outdoor air flow rate. This shows that 10 L/s (20 cfm) of outdoor air reduces health risk to a point of diminishing returns. Although a somewhat lower risk factor for involuntary risk from tobacco smoke would be desireable, a large increase in outdoor air flow rate would be needed for a significant reduction in risk. Thus, the recommendation for tobacco smoke control are not expected to change greatly.

A change in format, similar to the Nordic Standard, for presenting the recommended air flow rates is under consideration. The proposal is to use a building block approach. A basic ventilation rate for a space, independent of occupants, would be specified to control humidity, odors, etc. The requirements for metabolic needs, odor control, tobacco smoke control, etc. would be added to the basic building ventilation rate. Allowances could be made for expected ventilation effiency with various distribution systems and applications. Allowances would be made also for known internal contaminant sources such as office copying machines.

This approach would allow the designer to use whichever factors fit his particular application and design the ventilation to fit his particular needs.

The question of ventilation efficiency is also under study. It is known that mixing is often

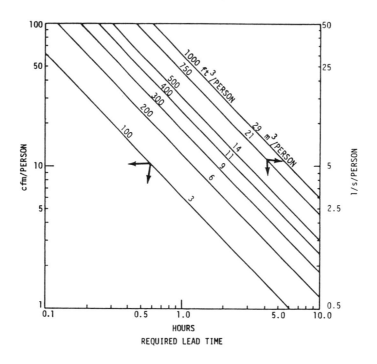

Figure 2 — Minimum ventilation time required before occupancy of space.

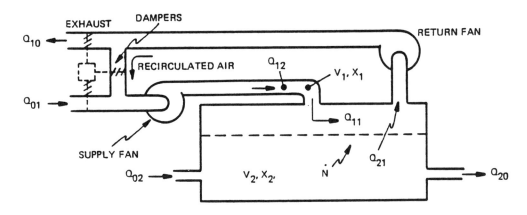

Figure 3 — Typical air distribution system.

imperfect. Figure 3 is a schematic of a typical air handling system.

Outdoor air at a flow rate Q_{01} is inducted, mixed with recirculated air and is delivered to the space though ceiling diffusers. Some of this may bypass the occupied level of the room and pass directly to the return system where a part of this return air $Q_{10\text{-}s}$ is exhausted. If s is the fraction of supply air that short circuits to the return system and r is the fraction of return air that is recirculated, it can be shown that the fraction of outdoor air inducted that reaches the occupied level (i.e., the ventilation efficiency) is given by equation 6.[9]

$$n_v = (Q_{01} - Q_{10\text{-}s})/Q_{01} = (1 - s)/(1 - rs) \quad (6)$$

Some limited data suggest that 50% ventilation efficiency may not be uncommon,[9] Variable Air Volume Systems appear to be particularly susceptible to the bypassing described here. The ventilation standard must address this problem.

Biological contaminants must be considered also. Air conditioning coils, drain pans and cooling towers can provide a place for growth of leagonella bacteria if maintenance is poor. Mold spoors can grow in damp areas and be circulated by a ventilation system. It is expected that some precautionary language will be added to the revised standard to cover biological contaminants.

Conclusions

Everyone abhors constraining regulations, but some are inevitable if society is to be adequately protected. The objective of the ventilation standard is to define a set of conditions which a designer can use to design a building ventilation system that will achieve an optimum balance among energy requirements, health, safety and comfort. Standard 62-1981 represented a step forward, but it needs explanation and clarification to be properly understood and accepted.

References

1. *Standards for Natural and Mechanical Ventilation.* ASHRAE Standard 62-73, ANSI B194.1-1977. The American Society of Heating, Refrigerating, and Air-Conditioning Engineers, 1791 Tullie Circle, NE, Atlanta, GA 30329.

2. *Ventilation for Acceptable Indoor Air Quality.* ASHRAE Standard 62-1981. The American Society of Heating, Refrigerating, and Air-Conditioning Engineers, 1791 Tullie Circle, NE, Atlanta, GA 30329.

3. Woods, J.E.: Ventilation, Health and Energy Consumption: A Status Report. *ASHRAE Journal,* p. 23 (July 1979).

4. *American Standards Building Requirements for Light and Ventilating — A53.1.* The American Standards Association (1946).

5. *Energy Conservation in New Building Design.* ASHRAE Standard 90-75. American Society of Heating, Refrigerating, and Air-Conditioning Engineers, Inc.

6. Brundrett, G.W.: *Ventilation Requirements in Rooms Occupied by Smokers — A Review.* Electricity Council Research Center, Pub. ECRC/M870. Capenhurst, Chester, England (December 1975).

7. Repace, J.: Report on Current Knowledge of Tobacco Smoke. *Engineering Foundation Conference on Management on Tightly Enclosed Spaces.* Engineering Foundation, United Engineering Trustees, Inc., J.E. Janssen Conference Chairman. Proceedings to be published by ASHRAE.

8. Janssen, J.E.: Air Mixing Efficiency. Presented at *Engineering Foundation Conference on Management of Atmospheres in Tightly Enclosed Spaces,* Santa Barbara, CA (October 1983). To be published by ASHRAE in proceeding of conference.

9. Janssen, J.E: Ventilation for Control of Indoor Air Quality: A Case Study. *Environment International* 8:487-496 (1982).

OFFICE BUILDING VENTILATION: ASSESSMENT

Tracer gas measurements of ventilation in occupied office spaces*

DAVID T. GRIMSRUD
Building Ventilation and Indoor Air Quality Program Applied Science Division, Lawrence Berkeley Laboratory, University of California

Introduction

Air quality within buildings depends upon two competing processes — the injection of a contaminant into the air from major pollutant sources within the building and the contaminants removal. This paper concentrates on the latter problem, contaminant removal.

Several removal processes can be considered for each particular contaminant present in the air. The process that is common to all contaminants is ventilation — contaminant control by diluting indoor concentrations by outdoor air. (This statement obviously must be qualified if the assumption that the outdoor air is relatively free of contamination is not true. In that case dilution control must include an initial cleaning of the outdoor ventilation air.)[1]

In this discussion we shall concentrate on the following two issues:

1. How much outside air is supplied to the space?
2. How is that outside air distributed throughout the space?

The latter question, in turn, has two parts: 1) How is the outside air distributed among zones in the building, and 2) how is it distributed within an individual zone?

Measurements of ventilation within an occupied space can be made in many ways; this paper discusses ventilation measurements using tracer gases. These techniques are not necessarily the best to tell us the total amount of outside air entering the building but they clearly have the potential to answer the second major ventilation issue, i.e., the distribution of the outside air throughout the space. For this reason both questions will be considered using the common measurement technique of tracer gas measurements as the reference procedure.

Tracer gas properties

There are several ways to think about how a tracer gas is used.[2-5] One particularly useful construct is to consider it a way to tag or to differentiate one parcel of air from another. For example, a tracer gas can differentiate the air within the building from the outside air. Because of this use, a tracer gas must possess specific properties:

1. It should not be a normal constituent of the air.
2. It should be a non-reactive gas so that its movement through the volume can be monitored.
3. It should be easily measureable at low concentrations.
4. It should be non-toxic and non-allergenic so that it can be used in occupied spaces.
5. It should have a molecular weight close to that of air so that the potential of stratification during mixing of pure tracer gas with the air are minimized.
6. It should be non-flammable.
7. It should be inexpensive.

No single tracer gas meets all the criteria listed above. Several studies have compared various tracer gases[6,7] and have shown their performance to be approximately equivalent. Nitrous oxide (N_2O) (commonly called laughing gas) is measureable in the parts per million (ppm) range using infrared absorption analyzers. It has been used widely as a tracer gas in the past, particularly in Europe. Its use in the United States has been

---

* This work was supported by the Assistant Secretary for Conservation and Renewable Energy, Office of Building Energy Research and Development, Building Systems Division of the U.S. Department of Energy under Contract No. DE-AC03-76SF00098.

reduced in recent years, however, because of potential harmful effects at trace concentrations.[8] It should not be used in occupied buildings.

Methane (CH_4) and ethane (C_2H_6) can both be measured at ppm concentrations using infrared analyzers. They each share the disadvantage of flammability, their lower flammability limits being 3%. If care is taken in mixing the gases into the volume they are suitable tracer gases. Another tracer that can be measured in the ppm level with an infrared analyzer is carbon dioxide (CO_2). Its background in outside air is of the order of 350 ppm; the occupants of the building are also sources of the gas. In some applications, described below, CO_2 is an attractive tracer for qualitative checks of ventilation rate and outside air distribution.

Each of the tracers listed are inconvenient to use in large building applications. A typical 1A cylinder of gas may contain 200 ft^3 of tracer. If we plan to inject tracer into a large building having a volume of 4,000,000 ft^3, a single injection to achieve a concentration of 100 ppm would require two cylinders of gas. Injection of such large volumes of gas is not only expensive but is also a logistical nightmare.

Tracer gases that can be analyzed using gas chromatograph instruments employing electron capture detectors are capable of detection at the parts per billion (ppb) to parts per trillion (ppt) level of concentrations. These concentrations clearly simplify the problem of injecting a sufficient amount of gas to obtain a detectable concentration. Sulfur hexafluoride (SF_6) is the tracer that has had the widest use in building applications. While it can be detected at ppm concentrations with infrared absorption instruments it can be detected at ppb to ppt concentrations using gas chromatographs. The molecular weight of SF_6 (146) means that some care must be used in initially mixing it with the air to be tagged. If premixed with dry nitrogen before injecting into the volume this problem can be avoided. Another group of gaseous compounds that can be detected with gas chromatographic equipment are the freons. These compounds offer the attractive possibility of multiple tracer use within the volume.

A special group of tracers that are used in a passive infiltration monitoring system developed at Brookhaven National Laboratory[9] are the perfluorocarbons. Because of the specialized procedures used in analyzing these tracers, they are not discussed further; however, the general measurement system is discussed further below.

Tracer gas use

General considerations

Discussions below of tracer gas techniques are all based on concentration measurements within a space. Changes in concentration are governed by a mass balance equation:

$$V (dC/dt) = F - QC \qquad (1)$$

where:

V = volume of the space (m^3)

C = volume concentration of the tracer gas (m^3/m^3)

F = injection rate of the tracer in the space (m^3/s)

Q = ventilation rate in the space (m^3/s)

The equation simply states that the rate of change of the amount of tracer within the space is equal to the difference between the amount of tracer added per unit time and the amount removed by ventilation. The equation assumes that ventilation is the only mechanism present that removes tracer gas from the space.

Concentration decay

The simplest application of Equation 1 is the case in which the tracer is mixed throughout the volume to achieve an initial uniform concentration, C_o. At this time (t_o) the injection is stopped. Substituting $F = 0$ into Equation 1 and integrating gives:

$$C(t) = C_o \exp \left[- (Q/V) (t - t_o) \right] \qquad (2)$$

where:

$C(t)$ = the tracer concentration at time t (m^3/m^3)

C_o = the tracer concentration at time t_o

and the other terms have been defined in Equation 1.

Equation 2 indicates that plotting the logarithm of C as a function of t will yield a straight line whose slope is given by $-Q/V$. Implicit in this solution are the assumptions that:

1. The ventilation rate, Q, is a constant over the time of measurement.
2. The volume that participates in the ventilation process is the physical volume of the space.
3. The space that is sampled is a single zone.

If assumption 1 or 3 is not true the decay will not be linear (when plotted on semi-log paper). If 2 is not true the decay will continue to be linear but the error in V will cause a corresponding error in the value of Q that is calculated from the slope.

A special case of concentration decay measurements of ventilation has been described by Penman (1980)[10] and Turiel and Rudy (1982).[11] Occupants of a building act as sources of CO_2 and cause the concentration of CO_2 within an occupied space to increase above the outdoor background concentration. If the occupancy suddenly decreases (e.g., at the end of the workday) we have a situation in which the building is charged with CO_2 while the injection rate from the occupants is zero. In this situation the concentration of CO_2 within the space will decay exponentially. This provides an opportunity to assess the ventilation of the space at a particular time each day. The only equipment requirement for this procedure is an analyzer to monitor the concentration of CO_2 and appropriate equipment to log the data produced. A variation of this technique will be described below.

Constant injection

If the injection rate of tracer gas into the space is held constant over time, the ventilation rate in the space can be calculated by solving Equation 1 for Q.

$$Q = (F/C) - [(V/C)(dC/dt)] \quad (3)$$

In steady state the second term on the right vanishes and the ventilation rate is given simply by the ratio of the injection rate to the concentration. How long must one wait for steady state to occur? The characteristic time for air mixing in the volume is of the order of V/Q where V is the volume of the space and Q is its ventilation rate. One must wait for times of the order of this characteristic time for the system to be governed by Equation 3. The procedure described is inherently an iterative process. Estimated solutions of Equation 3 are used to produce an appropriate time constant; these, in turn can be used to define the time one must wait so that Equation 3 will be valid.

An additional feature of Equation 3 that makes the technique attractive for use is the limited importance of the volume term in computing the ventilation rate. Whenever the concentration is changing slowly, the term $(V/C)(dC/dt)$ is small. Therefore, an uncertainty in V is not translated into an equal uncertainty in Q.

Long-term average concentrations

A special case of constant injection is the passive sampler system developed by Dietz and Cote.[12] It is useful to solve Equation 1 for the concentration, C, to better understand this measurement procedure.

$$C = (F/Q) - [(V/Q)(dC/dt)] \quad (4)$$

Taking time averages (denoted by the additional {}) of each of the terms in Equation 4 yields:

$$\{C\} = \{(F/Q)\} - \{[(V/Q)(dC/dt)]\}$$

or

$$\{C\} = F\{(1/Q)\} \quad (5)$$

Equation 5 is written assuming that the injection rate of the tracer, F, is constant in time. The second term on the right vanishes because the time average of a time derivative of a function is zero if the function is bounded. Since the function differentiated is the concentration, the time average is zero. The simplicity of this measurement system is reflected in Equation 5. The implementation of the system also makes the procedure attractive.

The tracer gases used, the perfluorocarbons, are detectable using the gas chromatographic system at Brookhaven National Laboratory at concentrations of parts per quadrillion (10^{15}). This means that a small container can be used as the source of tracer gas. The collector is a passive diffusion sampler that collects an air sample containing the tracer at a constant rate over the measurement period. At the conclusion of the measurement (encompassing times that range from one day to three months) the samplers are sent back to the laboratory for analysis. We emphasize that the procedure has not been validated yet in large buildings; several validation projects are in progress in residences. Finally, we again underscore the idea that the measurement result is an average concentration, that yields a ventilation rate that is likewise averaged over the measurement interval.

Constant concentration

The final procedure to discuss is the constant concentration procedure using a feedback machine. Referring again back to Equation 1, when tracer is injected into the volume at exactly the rate it is removed by ventilation, the concentration remains constant. When this occurs, the ventilation is given simply by the ratio of the injection rate and the concentration.

While this is a simple idea, it is difficult to achieve in practice. The problem is the inherent instability in the feedback process because of the finite mixing times involved in the ventilation process. If a sensor detects a decrease in concentration, the feedback machine will increase the injection rate. However, the concentration sensor will not detect an increase in the concentration until some finite time has passed to allow the gas to mix throughout the volume. This causes an inherent instability in the process and means that the feedback machine cannot be a simple device. The procedure has been used in several research settings but appears to be too complicated to expect to use as a routine measurement procedure.

Distribution of ventilation air

Two major effects complicate the measurement procedures using a tracer gas. The first is the multizone tracer gas problem. If two zones are coupled together (i.e., connected so that air can flow from one to the other) and tracer is injected into the first, the gas will leak into the second chamber as well as to the outside. That tracer is not lost; some may return to the original chamber. This two-way flow complicates the decay pattern one encounters in a normal tracer decay experiment and requires a more careful data interpretation to understand the flows involved. This problem was treated by Dick many years ago; this work has been expanded by Sinden.[13,14]

A second complication arises because ventilation air is not distributed uniformly within a single space. The study of this latter phenomenon, called ventilation efficiency, is an active area of research among ventilation engineers.[15-17] A useful way to think about this problem is to use the concept of the "age" of the air in a space that is being sampled. The term "age" is used in the same manner as it is used in common practice; the time since the air entered the room (birth) until present. The ventilation efficiency in the space is given by the ratio between the mean age of air in an ideal space in which the mixing is perfect and the mean age of the air in the actual room. Minimizing the time the air spends in the room, then, maximizes the ventilation efficiency and the ability of the ventilation air to dilute pollutant concentrations within the space. Ventilation efficiency becomes a "figure of merit" that can be used to judge various ventilation designs. Studies show, e.g., that ventilation efficiency in a space improves when a high supply is coupled with a low exhaust in a heating mode and a low supply is coupled with a high exhaust in a cooling mode. While this result may confirm intuitions about system design, the more important issue is the use of the concept of ventilation efficiency to better understand local variations in pollutant concentrations and resulting personal exposures. These techniques are being developed and hold much future promise. We must emphasize, however, that although they are promising, they do not exist today as a standard diagnostic technique. Until they do, field measurement procedures will have to rely on the experience of the field monitoring personnel to separate a situation having a large pollutant source from one having poor ventilation removal.

Summary

Tracer gases can be used to measure total ventilation rates within buildings, the coupling of flows between different zones in a building and the local distribution of ventilation air. Each of these is a useful procedure for diagnosing potential problems within a building. The most useful of these, the measurement of ventilation efficiency, is a topic of major current interest among research scientists and engineers; one can expect significant developments in monitoring techniques soon.

References

1. *ASHRAE Standard 62-1981, Ventilation for Acceptable Indoor Air Quality.* The American Society of Heating, Refrigerating, and Air-Conditioning Engineers, Inc., Atlanta (1981).
2. Hitchen, E.R. and C.B. Wilson: A Review of Experimental Techniques for the Investigation of Natural Ventilation in Buildings. *Bldg. Sci. 2*:59-82 (1967).

3. Jennings, B.H. and J.A. Armstrong: Ventilation Theory and Practice. *ASHRAE Trans.* 77:50-60 (1971).
4. Sherman, M.H., D.T. Grimsrud, P.E. Condon and B.V. Smith: Air Infiltration Measurement Techniques. *A.I.C. Conference Instrumentation and Measurement Techniques.* Windsor (1980).
5. Hunt, C.M.: Air Infiltration: A Review of Some Existing Measurement Techniques and Data. *Building Air Change Rate and Infiltration Measurements.* ASTM STP 719 (1980).
6. Grimsrud, D.T., A.N. Pearman, M.H. Sherman: An Intercomparison of Tracer Gases Used for Air Infiltration Measurements. *ASHRAE Trans. 86(I)*:258-267 (1980).
7. Bassett, M.R., C.Y. Shaw and R.G. Evans: An Appraisal of the Sulphur Hexafluoride Decay Technique for Measuring Air Infiltration Rates in Buildings. *ASHRAE Trans. 87(II)*:361-372 (1981).
8. *Occupational Exposure to Waste Anesthetic Gases and Vapors.* HEW Pub. No. (NIOSH) 77-140 (1977).
9. Dietz, R.N. and E.A. Cote: Air Infiltration Measurements in a House Using a Convenient Perfluorocarbon Tracer Technique. *Environ. Intl. 8*:419-434 (1982).
10. Penman, J.M.: An Experimental Determination of Ventilation Rate in Occupied Rooms Using Atmospheric Carbon Dioxide Concentration. *Bldg. Env. 15*:45-47 (1980).
11. Turiel, I. and J. Rudy: Use of occupant Generated CO_2 as a Ventilation Rate Indicator. *ASHRAE Trans. 88(I)*:197-210 (1982).
12. Dietz, R.N. and E.A. Cote: Op cit.
13. Dick, J.B.: Experimental Studies in Natural Ventilation of Houses. *J. Inst. Heat. Vent. Engineers 17*:420-466 (1949).
14. Sinden, F.W.: Multi-Chamber Theory of Air Infiltration. *Build. Env. 13*:21-29 (1978).
15. Sandberg, M. and M. Sjoberg: The Use of Moments for Assessing Air Quality in Ventilated Rooms. *Build. Env. 18*:181-198 (1983).
16. Freeman, J., R. Gale and M. Sandberg: The Efficiency of Ventilation in a Detached House. *Proceedings of the Third A.I.C. Conference, Energy Efficient Domestic Ventilation Systems for Achieving Acceptable Indoor Air Quality.* London (1982).
17. Skaret, E. and H.M. Mathisen: Ventilation Efficiency. *Environ. Intl. 8*:473-482 (1982).

Measurement of HVAC system performance

JAMES E. WOODS, Jr., Ph.D., P.E.
Senior Staff Scientist, Honeywell Physical Sciences Center, Bloomington, MN

System performance criteria are proposed in this paper by which the air distributed to and within occupied spaces may be evaluated. These criteria are expresssed in terms of thermal, air quality, acoustic and energy parameters. Methods of objectively determining the factors required to compare the operating conditions of the system with the performance are also described.

Introduction

Indoor air quality has become a major focal point within the last decade due to at least four factors: 1) energy conservation has led to reduced infiltration and ventilation in occupied spaces; 2) synthetic materials have been used more extensively; 3) effluents from indoor sources, such as tobacco smoke, copy machines, and aerosols have become ubiquitous; and 4) methods of detecting indoor pollutants at concentrations below those found in industrial facilities have become available. As a result, causal relationships and implications between concentrations of indoor contaminants, and health, comfort and productivity have been reported in local, national, and international publications.

These relationships have led to the following premises:

- As buildings become more energy efficient, they become less forgiving environmentally;
- Buildings which are controlled for environmental acceptability can operate energy efficiently; however,
- Buildings which are controlled for energy efficiency may not operate within environmentally acceptable criteria; thus,
- Energy efficiency is a necessary, but not sufficient, criterion for environmental acceptability.

Although indoor air quality has become a major concern, it has not yet been well-defined. For example, in the National Academy of Sciences' publication *Indoor Pollutants*,[1] methods of identifying, monitoring, and controlling indoor pollutants were addressed, but "Indoor Air Quality" remained undefined. For purposes of this discussion, the definition proposed for IAQ assessment programs in Iowa will be used:[2]

"The quality of the air in an enclosed space is defined as an indicator of how well the air satisfies the following conditions:

- *Thermal conditions of the air (i.e., dry-bulb temperature, relative humidity and velocity) must be adequate to provide thermal acceptability for the occupants as defined by ASHRAE Standard 55-1981.[3] Effects of mean radiant temperature, thermal resistance of clothing, and the occupants' activity levels must be considered in this evaluation.*

- *Concentrations of oxygen and carbon dioxide must be within acceptable ranges to allow normal functioning of the respiratory system.*

- *Concentrations of gases, vapors, and aerosols should be below levels that can have deleterious effects, or that can be perceived as objectionable by the occupants."*

Basic control strategies

Conventional control strategies are based on the assumption that the air in the occupied spaces is uniformly mixed. Although this assumption is valid in some applications, poor air quality is often correlated with non-uniformly mixed or stratified air which results in exposures to concentrations that exceed acceptable levels.

One-compartment models

A steady-state, one-compartment model of an indoor air quality system is shown in Figure 1. The relationship of the three most common methods of control (i.e., source control, $\dot{N}$; removal control, $\dot{E}$; and dilution control, $\dot{V}$) is shown. In this model, source control may be represented by isolation, product substitution or local exhaust; dilution

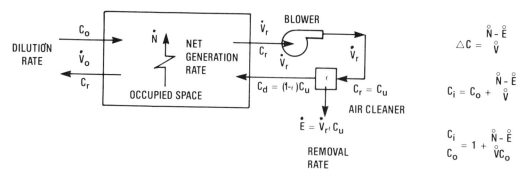

Figure 1 — One-compartment, uniformly-mixed, steady-state model for indoor quality control.

control may be represented by infiltration, natural ventilation, or mechanical ventilation; and removal control may be represented by fan-filter modules, clean benches, or central forced air systems with recirculated air.

This model indicates that the indoor air concentration of a contaminant, C_r, should be expected to exceed the outdoor air concentration C_o, unless the removal rate, $\dot{E}$, exceeds the net generation rate, $\dot{N}$, or the dilution rate, $\dot{V}$, is infinitely large. Although the latter control method is impractical, the former method is commonly used for applications such as clean rooms. This relationship also indicates a reasonable control strategy: to achieve an acceptable ΔC, first attempt to reduce the net generation rate, then apply techniques that will increase the removal rate, then use techniques that will increase the dilution rate as necessary.

Another important concept indicated by this model is that only source control methods can eliminate risk of occupant exposure; both removal and dilution control methods may cause mixing of the contaminant with the air in the occupied space. The latter control methods may be considered as *ventilation control*, and when combined with source control, the three strategies may be considered as *air quality control*. As shown in Table I, this concept is compatible with the definitions of "General Ventilation" and "Dilution Ventilation" given by ACGIH,[4] and the definition of "Ventilation" given by ASHRAE.[5] However, these definitions are significantly different than the definition of ventilation commonly assumed by physiologists and other life-scientists, also shown in Table I. The fundamental difference is that the engineer considers ventilation air as that entering the occupied space (i.e., room ventilation), while the physiologist considers ventilation air as that being respired (i.e., respiratory ventilation).

Two-compartment models

To couple the concepts of room and respiratory ventilation, a two-compartment model may be considered. As shown in Figure 2, a steady-state ratio between the room and respiratory ventilation rates, $(\dot{V}_o/\dot{V}_a)$, can be expressed as the ratio between the contaminant concentrations differences of the expired and inspired (i.e., room) air and the room and outdoor air. Table II indicates two examples of ventilation ratios; one for carbon dioxide and one for oxygen. Note that for the case where the steady-state room air concentration of CO_2 is assumed to be maintained at 0.25%,[5] a ventilation ratio of approximately 17 is required;

TABLE I
Physiological and Engineering Definitions of "Ventilation"

Physiological definition:	"The inspiration of *Fresh* air followed by the expiration of some alveolar gas."
Engineering definitions:	
ASHRAE[5]	"The process of supplying and removing air by natural or mechanical means to and from any space. Such air may or may not be conditioned."
ACGIH[4]	"*General Ventilation* refers to the removal (or supply) of air in a general area, room or building for the purpose of comfort control."
	"*Dilution Ventilation* refers to the dilution of contaminated air in a general area, room or building for the purpose of health hazard or nuisance control."

that is 17 times more air is required to maintain the acceptable room concentration than required to maintain a normal concentration in the expired air (i.e., 3.8%). Also note that a ventilation ratio of 9 is required to maintain an acceptable room air concentration of oxygen. Therefore, as removal techniques for CO_2 and O_2 are not commercially practical at this time, this example indicates that maintenance of a room concentration of 0.25% CO_2 will require nearly twice as much dilution air as maintenance of 20.5% O_2.

These relationships are particularly significant, as several values for the "acceptable level" of CO_2 have been promulgated: the Scandinavian countries are currently recommending that concentrations of 0.07-0.08% not be exceeded;[6] the World Health Organization has recommended that concentrations of 0.1% not be exceeded;[7] the Japanese indoor standard is now set at 0.10%;[8] ASHRAE has recommended that concentrations of 0.25% not be exceeded on a continuous basis;[5] OSHA has set a time-weighted average of 0.5%;[9] the US Navy has set a maximum long-term exposure level of 0.5%;[10] and NASA has set a maximum long-term exposure level at 1.0%.[11] While the latter three values have been set for direct exposure to CO_2, it is probable that the former four values correlate to other contaminants that also exist with the specified levels of CO_2, and that these other contaminants are the sources of objectionable indoor conditions.

Another form of a two-compartment model of an occupied space has been reported by Janssen et al.[12] As shown in Figures 3 and 4, this model indicates that "Ventilation Efficiency" for the occupied space can be expressed as a function of the percent of recirculated air, R, and a stratification factor, S. In this model, S is defined as the ratio of the initial and steady-state decay rates $(I_0 - I_\infty)$ divided by the initial decay rate (I_0). Of importance in this model, the stratification factor can be associated with the location of the supply and return air terminal devices, and the recirculation percentage can be associated with the ventilation control system.

Two-compartment and multi-compartment models may also be used to assess air quality control between rooms. One such model, shown in Figure 5, has been used as a basis for describing "Ventilation Effectiveness" and "Relative Exposure

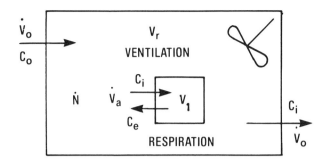

VENTILATION: $V_r \dfrac{d(C_i)}{dt} = \overset{\circ}{N} - \overset{\circ}{V_o}(C_i - C_o)$

Steady state: $\overset{\circ}{V_o}(C_i - C_o) = \overset{\circ}{N}$

RESPIRATION: $V_1 \dfrac{d(C_e)}{dt} = \overset{\circ}{N} - \overset{\circ}{V_a}(C_e - C_i)$

Steady state: $\overset{\circ}{V_a}(C_e - C_i) = \overset{\circ}{N}$

COMBINED STEADY STATE EQUATIONS:

$$\dfrac{\overset{\circ}{V_o}}{\overset{\circ}{V_a}} = \dfrac{C_e - C_i}{C_i - C_o}$$

Figure 2 — Two-compartment model of the coupling between room and respiratory ventilation. In steady-state, the ratio of room and respiratory ventilation rates is defined as the "ventilation ratio."

Index" in a forced air system.[13] In this model, the ventilation effectiveness was described as the ratio of the areas under the decay curves of a reference room and other rooms:

$$\eta = \int_0^\infty C_R dt / \int_0^\infty C_i dt \quad \bigg| \quad C_i = \text{uniform}$$

Results of data obtained in a three-bedroom house are shown in Figure 6 for the case when the forced air fan was energized, and in Figure 7 when the forced air fan was de-energized. These results indicate that, although uniform mixing may exist when the forced air system is operational, significant differences between rooms may develop during light-load conditions when thermostatic

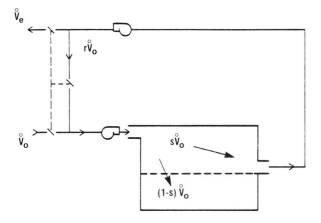

Figure 3 — Two-compartment model of a room with stratification of the indoor air.[12]

control may cause the forced air system to be de-energized for extended periods.

The extent of the differences that may exist between rooms is further indicated in Figure 8. Rather than challenging the system with the same concentration of tracer gas in each zone, if the tracer is introduced in only one room (e.g., the room in which the suspected contaminant source exists), a more powerful evaluator of non-uniform mixing may be expressed as the Relative Exposure Index:[13]

$$REI = \int_0^\infty C_i dt / \int_0^\infty C_R dt \quad \Big| \quad C_i = local$$

Figure 8 indicates the results of challenging the system with a tracer gas in the living room. Note that, with the fan de-energized in this case, concentrations in all three bedrooms exceeded the concentration in the living room after about five hours. These results are especially important for situations where a local heating device, such as a ventless space heater, is used and causes the thermostat to de-energize the fan in the central forced air system.

Results from the two- and multi-compartment models and supporting data lead to an important consideration for assessing HVAC system performance:

Methods of room air distribution may be as important to indoor air quality control as methods of system air distribution.

System performance criteria

Because of the relatively similar importance of roof air and system air distribution, both factors should be evaluated with equal thoroughness. Criteria with which to evaluate these factors can be expressed in terms of thermal, air quality, acoustic, and energy parameters. For clarity, terminology associated with heating, ventilating, and air conditioning of occupied spaces has been adopted from ASHRAE Standard 62-1981. Many of the relevant terms are shown schematically in Figure 9.

Room air distribution

Air movements from the supply air terminal units to the occupants, and then to return or exhaust air terminal units should be evaluated for its ability to provide acceptable thermal conditions, acceptable air for respiration and suppression of airborne contamination, and acceptable acoustic conditions.

Acceptable thermal conditions may be defined as those that comply with ASHRAE Standard 55-1981 at the location of the occupants (i.e., near their micro-environments). The effectiveness of distributing the supply air to the occupants's locations may be evaluated in terms of the Air Diffusion Performance Index (ADPI).[14] This concept, shown in Figure 10, is based on the ability of

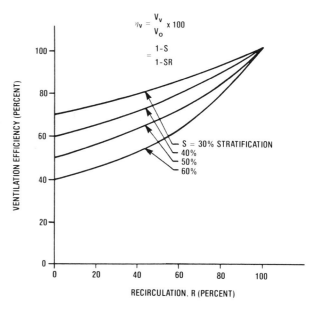

Figure 4 — Ventilation efficiency as a function of recirculation percentage in a room with stratification of the indoor air.[12]

the supply air to remove sensible heating loads from the occupied space. Although it was originally intended as an evaluator for the selection and location of supply air terminal units, it can also serve as a reasonable index for evaluating the uniformity of the thermal conditions of the air throughout the room for either heating or cooling. Terminology that is used in determining the ADPI is shown in Figure 11 and Table III.

For purposes of providing an evaluator of air movement to provide acceptable thermal conditions within an occupied room, a minimum ADPI of 75% is proposed. Ratios of isothermal throws of supply terminal devices to room characteristics lengths (T_{50}/L), which are expected to provide a minimum ADPI of 75%, are shown in Table IV. Curves relating the T_{50}/L ratio of various supply air terminal devices to ADPI as a function of the sensible heat load in the occupied space are available.[14,15] One such curve for circular ceiling diffusers is shown in Figure 12. The horizontal dashed lines at Q/A = 80 and 20 Btu/hr-ft² represent an improvement of ADPI from 75 to 92% that could be expected as the thermal load decreases, if the occupied space were controlled by a constant air volume (CAV) system (i.e., $T_{50}/L = 1.2$). However, if the occupied space were controlled by

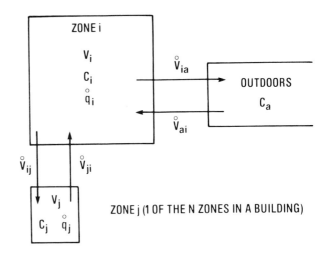

Figure 5 — Multi-compartment model of indoor air quality in an occupied space as affected by outdoor air and other zones within the building.[13]

a variable air volume (VAV) system, Figure 12 indicates that the ADPI would increase to 95%, or an increment of 3% over the CAV could be expected as the sensible load decreases. Moreover, the corresponding decrease in T_{50}/L from 1.2 to 0.8 represents a decrease in required fan horsepower, which also makes VAV systems attractive from an energy savings viewpoint. A disadvantage of the

TABLE II
Room and Respiratory Ventilation of a Normal Healthy Adult, Indoors and Outdoors

Function	Symbol	Units	Outdoors		Indoors	
			O_2	CO_2 (RQ=0.83)	O_2	CO_2 (RQ=0.83)
Generation rate	N	L/min	-0.36	0.03	-0.36	0.30
		m³/hr	-0.022	0.018	-0.022	0.018
Concentration of inspired (room) air	C_i	%		0.04[a]	20.5[b]	0.25[c]
		m³/m³ air	0.029	0.004	0.205	0.0025
Concentration of normally expired air	C_e	%	16.8	3.8	16.8	3.8
		m³/m³ air	0.168	0.038	0.168	0.038
Concentration of typical outdoor air	C_o	%	20.9	0.04	20.9	0.04
		m³/m³ air	0.209	0.0004	0.209	0.0004
Respiratory ventilation rate:	V_a	m³/hr	0.54	0.48	0.59	0.51
$V_a = N/(C_e - C_i)$		cfm	0.32	0.28	0.35	0.30
Room ventilation rate:	V_o	m³/hr	—	—	5.5	8.6
$V_o = N/(C_i - C_o)$		cfm	—	—	3.2	5.1
Ventilation ration:	V_a/V_o		—	—	9.3	16.9

[a]Typical outdoor concentrations. [b]Typical indoor concentrations. [c]ASHRAE recommended limit.

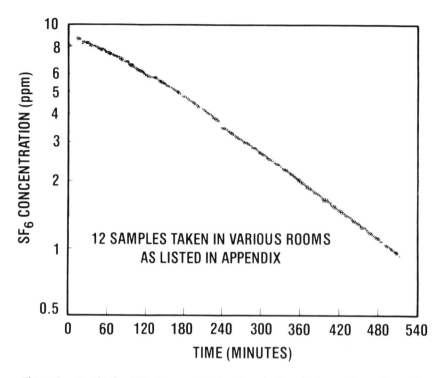

Figure 6 — Ventilation Effectiveness of 12 locations in three bedroom Energy Research House at Iowa State University. Results shown were obtained after all zones were equalized at the initial tracer gas concentration; the central fan was remained energized during the decay period.[13]

VAV systems is that current control stategies are insensitive to changes in latent heat loads or ventilation requirements that occur from changes in occupancy density. Thus, the ADPI should be considered as a convenient, but insufficient, criterion for evaluating the performance of a room air distribution system.

Acceptable air quality conditions may be defined as those that comply with ASHRAE Standard 62-1981 at the location of the occupants (i.e., near their micro-environments). Two procedures for providing acceptable air quality are specified in that standard: a "Ventilation Rate Procedure" and an "Air Quality Procedure." These procedures are intended to provide acceptable conditions at the occupants' locations, but explicit methods for evaluating compliance have not been specified for either procedure. The "Ventilation Rate Procedure" specifies volumetric flow rates of outdoor, or filtered recirculated air, that should provide acceptable air quality, but methods of assuring that the ventilation air supplied to the room is also adequately distributed to the occupants are not specified. The "Air Quality Procedure" specifies concentrations of some contaminants that should not be exceeded in the occupied space, but methods of assuring that the concentrations are not exceeded are not specified. The ADPI can be used to evaluate the thermal uniformity of air mixing in the occupied spaces, but as previously indicated, it is insufficient as an overall evaluator of the room air distribution. For example, the ADPI does not provide for an evaluation of the amount of supply air that can bypass the occupied space because of short-circuiting of air from the supply to the return air terminal devices. However, as indicated in Figures 3 and 4, Janssen's "Within Room" Ventilation Efficiency[12] does provide an evaluator of the distribution of the supply air to the occupants.

For purposes of providing an evaluator of air movement to provide acceptable air quality conditions within an occupied room, a minimum "Within Room" Ventilation Efficiency of 80% is proposed.

Acceptable acoustic conditions within occupied spaces probably have more variance than either thermal or air quality conditions. At this time, no standards for acceptable acoustic conditions exist, but ranges from 25-30 to 40-45 dB (Noise Criteria or Room Criteria Curves) for 12 functional areas have been recommended by ASHRAE.[16] Of particular concern for room air distribution systems is the sound generated by the supply, return, and exhaust terminal units as air flows past them.

For purposes of providing an evaluator for acceptable acoustic performance of room air terminal units, the sound pressure level from the terminal units and associated volume dampers shall not exceed the specified Room Criteria range for the functional area when the air supply rate is at its maximum.

System air distribution

To evaluate the performance of a system, it is necessary not only to consider each component of the system but to consider the effects of their coupling. In some cases, the system will consist of a pre-packaged unit coupled to the occupied space by ductwork, piping, and a set of controls. In other cases, the system will consist of a "built-up" set of components that are coupled to the occupied space by ductwork, piping, and a set of controls. In both cases, the system will function in response to a control strategy. Because of the uniqueness of the relationship of each system, performance evaluation criteria must be site-specific, and relate to both design and off-design (i.e., part-load) conditions. Thus, the systems within the building should be evaluated individually and collectively for their ability to provide the required thermal, air quality, and acoustic conditions to the occupied spaces energy efficiently and cost effectively.

To meet the thermal requirements within the occupied spaces while providing for energy efficiency, either the enthalpy or the flow rate of the

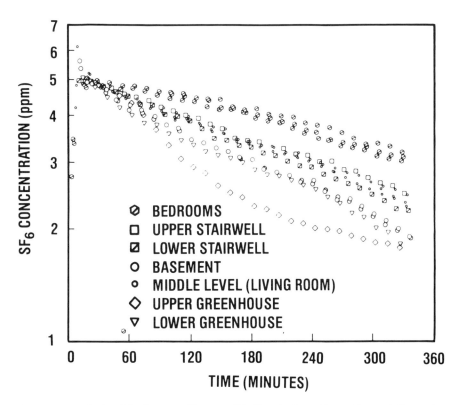

Figure 7 — Ventilation Effectiveness of 7 zones of the three bedroom Energy Research House at Iowa State University. Results shown were obtained after all zones were equalized at the same initial tracer gas concentration; the central fan remained de-energized during the decay period.[13]

TABLE III
Characteristic Lengths, L, for Terminal Devices*

Terminal device	Characteristic Lengths, L
High sidewall grilles	Distance from grille to opposite wall or to intersecting air jet plus ceiling height minus 8 feet.
Ceiling diffusers Slot diffusers Light troffer diffusers	Distance from center of diffuser to closest wall, or to intersecting air jet plus ceiling height minus 8 feet.
Sill grille	Length of room in direction of jet flow.

*Adapted from Nevins and Miller,[14] and Nevins.[15]

supply air must vary in response to changes in the thermal loads. For constant volume systems (CAV), the enthalpy will change and the supply air volumetric flow rate will remain approximately constant with two exceptions: two-position thermostatic control will cause the supply air flow rate to be de-activated for various periods as a function of thermal load, and blower speed may be adjusted for different flow rates during heating and cooling seasons. For variable air volume systems (VAV), the supply air volumetric flow and the enthalpy will change. The amount of change in the supply air conditions is dictated by two factors: provision of acceptable thermal conditions in the occupied spaces, and the energy that must be consumed to meet those conditions.

Criteria for evaluating the thermal performance of the system are proposed as follows:

- The supply air volumetric flow rate shall be sufficient to maintain a minimum ADPI of 75% during occupied periods of the zones.

- The dry-bulb temperature of the supply air shall be sufficient to maintain the following conditions specified in ASHRAE Standard 55-1981: 1) acceptable ranges of operative temperature for winter and summer conditions; 2) a maximum vertical temperature gradient in the occupied space of 3°K (5°F); and 3) a maximum rate of temperature change in the occupied space of 2.2°K/hr (4°F/hr), if the peak variation in operative temperature exceeds 1.1°K during the control cycle.

- The dew-point temperature of the supply air shall be sufficient to maintain an acceptable dew-point temperature range in the occupied space as defined in ASHRAE Standard 55-1981.

To meet the air quality requirements within the occupied spaces while providing for energy efficiency, either the dilution rate, $\dot{V}$, or the removal rate, $\dot{E}$, (see Figure 1) must vary in response to changes in the generation rate, $\dot{N}$, and the outdoor concentration C_o. Because closed-loop air quality control systems are not generally employed in HVAC systems, acceptable air quality conditions within occupied spaces are usually maintained by providing at least a minimum amount of outdoor air for dilution through the system, as shown in Figure 9. Also, as shown in Figure 9, a percentage of the return air from the occupied spaces is often recirculated and mixed with the outdoor air for energy efficient thermal control. If the recirculated or mixed air is properly cleaned or filtered, this mixed air may also serve for removal control. For CAV systems, the percentage of outdoor air is usually minimized at winter and summer design conditions, and at part-load conditions, the increase in the percentage of outdoor is controlled to reduce energy consumption to provide acceptable thermal conditions. Except when two-position thermostatic control of the occupied spaces is employed, CAV systems will normally provide sufficient ventilation for variations in occupancy or process loads. For VAV systems, the minimum percentage of outdoor is usually sufficient for ventilation at design conditions because

TABLE IV
Range of Acceptable Throw Characteristics, T_{50}/L, for Minimum ADPI Values of 75%*

Terminal Device	Room Sensible Load (Btu/hr-ft^2)	T_{50}/L Range
High sidewall grilles	< 40	1.2 to 2.0
Circular ceiling diffusers	< 70	0.7 to 1.2
Ceiling slot diffusers	≤ 80	0.3 to 0.7**
Light troffer diffusers	≤ 60	< 3.0
Perforated and louvered ceiling diffusers	< 50	1.0 to 3.0
Sill grilles		
Vanes straight	< 50	1.3 to 1.7
Vanes spread	< 60	0.6 to 1.7

*Adapted from Nevins.[15]
**for T_{100}/L

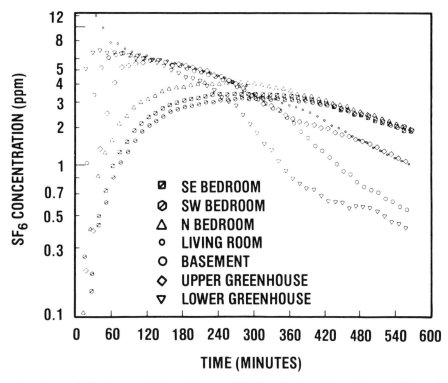

Figure 8 — Relative Exposure Index of 7 zones of the three bedroom Energy Research House at Iowa State University. Results shown were obtained after tracer gas was introduced into the living room; the central fan remained de-energized during the period of decay.[13]

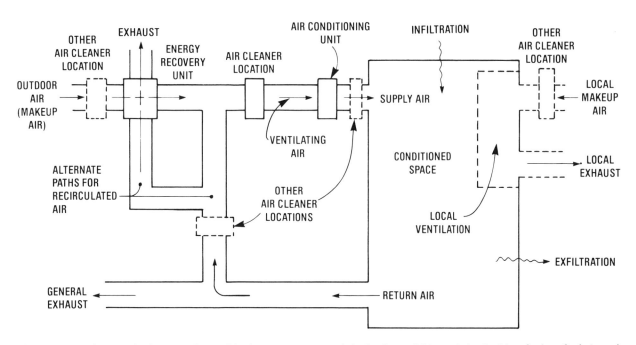

Figure 9 — Heating, Ventilating, and Air Conditioning system schematic indicating variables and standard terminology for indoor air quality control.[5]

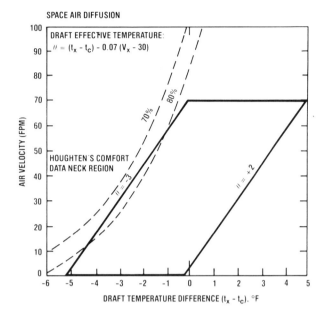

Figure 10 — Representation of the Air Diffusion Performance Index (ADPI), after Nevins and Miller,[14] and Nevins.[15] ADPI is the percentage of total points measured in an occupied space that meet the criteria: $-3 \leq \phi \leq +2; 0 < V \leq 70$.

the supply air flow rate is maximized. However, at part-load thermal conditions when the supply air flow rate is reduced, an increase in the percentage of outdoor air may be necessary to provide acceptable air quality conditions in the occupied spaces.

Criteria for evaluating the air quality performance of the system are proposed as follows:

- The quality of the outdoor air to be used for ventilation should be in compliance with ASHRAE Standard 62-1981, and sufficiently clean to provide dilution control of indoor contaminants that are not removed by other means.

- If acceptable indoor air quality is to be controlled by dilution only, the volumetric flow rate of outdoor air provided by the system should be in compliance with the "per person" requirements in ASHRAE Standard 62-1981 for "smoking" and "non-smoking" spaces during occupied conditions.

- If acceptable indoor air quality is to be controlled by a combination of dilution and removal, the volumetric flow rate of outdoor air can often be reduced. The revised volumetric flow rate of outdoor air may be determined as shown in Figure 13. For cases where the acceptable indoor concentrations exceed outdoor conditions by a substantial factor, ASHRAE Standard 62-1981 offers a simplified and more conservative method of determining the required amount of compensating recirculated air:

$$V_r = (V_o - V_m)/\epsilon$$

where V_m is an arbitrary reduced outdoor air rate and ϵ is the air cleaner efficiency rated in the same terms as the contaminant concentrations.

- A Zone Ventilation Index, (ZVI), has been defined as the ratio of the outdoor air percentage of supply air provided by the HVAC system to the outdoor air percentage of supply air required in an occupied space.[17] If the ZVI for any zone is less than one, the outdoor air percentage of supply air provided by the HVAC system is insufficient. If the ZVIs for all zones are greater than one, more than sufficient outdoor air is supplied by the HVAC system, and if thermal conditions are not compatible with mixed air control energy savings, the percentage of outdoor provided by the system may be decreased.

- To evaluate potential cross-contamination between rooms within a zone and between

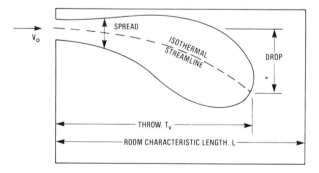

Figure 11 — Nomenclature associated with room air distribution.[14,15] *Throw:* distance air stream travels until specified terminal velocity is obtained. *Spead:* horizontal or vertical divergence of air stream after it leaves the diffuser. *Drop:* the vertical distance travelled by the streamline at its point of throw. *Room characteristic length:* that dimension of a room by which the room air diffusion can be described.

zones of a system, the Relative Exposure Index[13] should not exceed 1.2 when the zone containing the largest potential contaminant generation rate, N, is challenged by a tracer gas.

To meet the acoustic requirements within the occupied spaces while providing for energy efficiency, sources of noise generation within the HVAC system should be identified and supressed. In addition to the room terminal devices, sources of noise within the HVAC system include the supply, return and exhaust fans; fittings, branch take-offs, dampers, heat exchangers, mixing boxes, and other elements within the supply and return ductwork.

Criteria for evaluating the acoustic performance of the system are proposed as follows:

- Sound transmitted by components of the HVAC system to the occupied spaces through direct (i.e., through the ductwork) or indirect (i.e., through building materials) pathways should result in sound pressure levels that are 3-5 dB less than the appropriate Room Criteria for each functional area.

- Vibration transmitted by components of the HVAC system to the occupied spaces through structural pathways should not exceed the "Maximum Allowable Vibration Levels" recommended by ASHRAE[16] for various rotating and reciprocating equipment.

Two factors that can be used to evaluate the energy efficiency of the HVAC system are the energy requirements to provide the acceptable environmental conditions within the occupied spaces, and the energy consumed by the system to meet the energy requirements. The ratio of the sum of the energy requirements for all occupied spaces served by the HVAC system to the energy consumed by that system may be defined as the system energy efficiency. The primary system inefficiencies are caused by three types of parasitic losses: 1) air leakage through seams and connections in the ductwork, 2) excessive pressure drops imposed on the system, and 3) excessive capacities of the air handling systems.

Criteria for evaluating the energy performance of the system are proposed as follows:

- Air leakage between the system air handling units (e.g., supply, return, and exhaust air fans) and the occupied should be less than 10% of the air flow required to be delivered.

- The total static pressure drop of the system should not exceed that resulting from "Air Transport Factors" shown in Table V.[17]

- The fan efficiency ratio, defined as the fan efficiency at operating conditions divided by the maximum efficiency obtainable by the fan, should not be less than 0.9.[17]

- The overall energy efficiency of the system should not be less than 70%.

Performance evaluation

Objective methods of measuring and evaluating the room air and system air distribution are available. However, due to constraints such as costs and technical training, these methods are not routinely employed when assessing the performance of a system. Rather, system performances

TABLE V
Recommended Limits of Air Transport Factors for Primary Air Handling Equipment

Air Handling Capacity (cfm)	Air Transport Factor* (ATF)	Approximate Total Static Pressure** (in H_2O)
< 5000	12.0	2.7
5000 < 25,000	10.0	3.1
25,000 ≤ 100,000	8.0	5.0
< 100,000	4.0	10.0

* ATF = Space Sensible Heat Removal/(Supply Return Fans Power Input); expressed in consistent units

** Based on $\Delta t = 20°F$

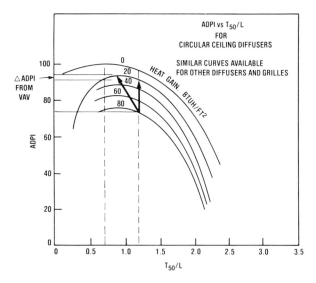

Figure 12 — Air Diffusion Performance Index, ADPI, as a function of the throw ratio, T_{50}/L, and heat gain flux, Q/A. Arrows indicate the increase in ADPI expected due to reduction in heat gain in a constant air volume system, CAV, and a variable air volume system, VAV.

are commonly evaluated by subjective and often superficial methods. As a result, occupant complaints of discomfort or sickness have occurred, needlessly. In the remainder of this paper, methods are proposed for obtaining thermal, air quality, acoustic, and energy measures, for comparing them to evaluation criteria, and for obtaining an overall performance evaluation of the system.

Room air distribution

Air movement within the occupied space may be evaluated for thermal acceptability by comparing the ADPI of the room with evaluation criterion of 75%. The ADPI for the room may be determined as follows:[17]

1. Measure the volumetric air flow rates and dry-bulb temperatures at all terminal devices in the room.
2. For the flow rates measured, determine the expected isothermal throws to a terminal velocity (e.g., 0.25 m/s or 50 fpm), T_v, of the terminal devices from catalog or manufacturer's data.
3. Determine the characteristic length (L) for each terminal device from Table III and calculate the parameter T_v/L.
4. From the total room supply air flow rated V_s (cfm), the average supply and return air dry-bulb temperatures, T_s and T_r (°F), and the floor area of the occupied space, A_2 (ft^2), determine the normalized room sensible heat gain, Q/A (Btu/hr-ft^2):

$$Q/A = [1.1 V_s (T_r - T_s)]/A$$

5. Compare results of points number 3 and 4 above with acceptable ranges of T_v/L in Table IV.

Air movement within the occupied space may be evaluated for acceptable air quality by comparing the measured "Within Room" ventilation efficiency with the evaluation criterion of 80%. The "Within Room" ventilation efficiency for a room may be determined as follows:[12,17]

1. Inject a tracer gas (e.g., SF_6, CH_4) into the supply air duct to the room and measure the decay in concentration in the return air duct from the room.
2. Determine the initial decay rate, I_o, during time $(t_i - t_o)$, and the final decay rate, I_∞, during time $(t_f - t_i)$ by least squares fits:

$$I_o = (\ln C_o/C_i)/(t_i - t_o)$$

and

$$I_\infty = (\ln C_i/C_f)/(t_f - t_i)$$

3. Determine the stratification factor, S:

$$S = (I_o - I_\infty)/I_o$$

4. Determine the outdoor air percentage of supply air (i.e., the system ventilation ratio, SVR) and the system recircuation percentage, R, from measured dry-bulb temperatures of the outdoor air, T_o, the recirculation air, T_r and the mixed air, T_m:

$$SVR = V_o/V_m = (T_r - T_m)/(T_r - T_o)$$

$$R = (1 - SVR) \times 100\%$$

5. Determine the "Within Room" ventilation efficiency, η, from Figure 4.

Room air terminal devices may be evaluated for acoustic acceptability by comparing catalog or manufacturer's data for the devices with the measured sound pressure levels of the room and the recommended Room Criteria:[16]

1. Measure the sound pressure levels of the room at the eight octave bands defined in the Room Criteria Curves (i.e., 31.5 to 4000 Hz).
2. Compare the measured values to the recommended Room Criteria for the appropriate functional area.
3. Obtain sound pressure ratings for the terminal devices from catalog or manufacturer's data.
4. Add incremental SPL of 2 dB for each 10% increase of air velocity above catalog rating.
5. Add incremental SPL of 7 dB for each doubling of throttled pressure drop at the device.
6. Add incremental SPL for N multiple devices:

$$SPL = 10 \log N$$

System air distribution

The supply air of the system may be evaluated for thermal acceptability by comparing the measured values of the volumetric flow rate, dry-bulb temperature, and dew-point temperature to those required to maintain acceptable ADPIs, operative temperatures, and dew-point temperatures in all occupied zones of the system:

1. The volumetric flow rate of the supply air should be determined by obtaining a standard pitot tube traverse in a horizontal section of duct at least 10 diameters downstream of the fan discharge and upstream of the first branch take-off.
2. The dry-bulb and dew-point temperatures should be measured at the same location at which the pitot traverse is obtained.

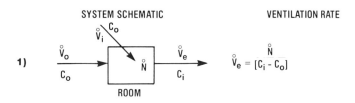

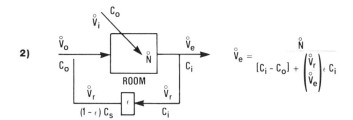

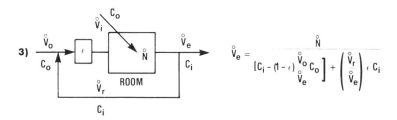

Figure 13 — One-comparment, uniformly-mixed, steady-state representations of three ventilation systems: 1) room ventilation provided only by outdoor air, V_o and V_i (100% outdoor air system); 2) room ventilation provided by outdoor air and by treated recirculated air (typical of residential systems); and 3) room ventilation provided by treated outdoor and recirculated air (typical of commercial systems). (Defined ventilation ratios: $\dot{V}_e = \dot{V}_o + \dot{V}_i$; $V_s = V_r + V_o$.)

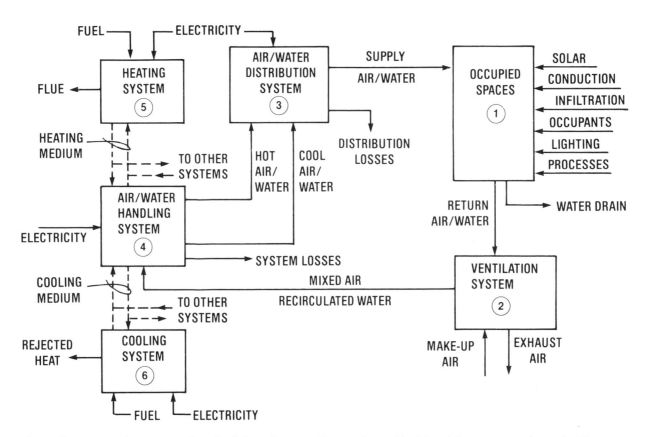

Figure 14 — Schematic representation of building subsystems that may be used in determining energy requirements, ER, energy consumption rates, EC, and energy efficiency, EE, of a building.[17]

The supply air of the system may be evaluated for acceptable air quality by comparing the measured values of the outdoor concentrations, the air cleaner efficiency, and volumetric flow rates with the evaluation criteria:

1. If the indoor air quality is to be controlled by the "Ventilation Rate Procedure" in ASHRAE Standard 62-1981, the quality of the outdoor air should be evaluated for compliance by comparing locally obtained data with Tables 1 and 2 in the Standard. If local records are not available, or if specific contaminants may be present, on-site measurements of the concentrations in the make-up air intakes should be obtained.

2. If the indoor air quality is to be controlled by the "Ventilation Rate Procedure" and recirculation is not employed, the measured volumetric supply air flow rate of the system should be compared to the product of the required rates per person in "smoking" and "non-smoking" spaces, as shown in Table 3 of Standard 62-1981, and the number of occupants in each of those spaces.

3. If the indoor air quality is to be controlled by the "Ventilation Rate Procedure" by a combination of removal and dilution, the efficiency of the air cleaner, ϵ, the air flow rate through the air cleaner, and the flow rate of outdoor should be determined and compared with appropriate values in ASHRAE Standard 62-1981. If efficiency data are not available from the manufacturer, an on-site estimation should be made by measuring the contaminant concentrations up-stream, C_u, and down-stream, C_d, of the air cleaner, and calculating the contaminant penetration, P, and the efficiency ϵ:

$$P = C_d/C_u$$
$$\epsilon = 1 - P$$

4. The Zone Ventilation Ratio, ZVI, may be calculated by determining the outdoor air percentage of the system' supply air (i.e., system ventilation ratio, SVR) and dividing by the required outdoor air percentage of the zone supply air (i.e., zone ventilation ratio, ZVR). The SVR may be determined from:

$$SVR = (T_r - T_m)/(T_r - T_o)$$

And the ZVR may be determined by calculating the required outdoor per person from ASHRAE Standard 62-1981, multiplying by the number of people in the zone and dividing by the supply air flow rate to the zone.

5. The Relative Exposure Index may be determined by injecting a tracer gas into the air within the occupied space suspected of having the primary source of contamination. The response curves, Figure 8, are then determined by sensing the tracer gas in the occupied spaces where exposure is of concern. The REI is then determined by step-wise integration of the response curves and division of the result from occupied space, i, by result from the reference (i.e., source space), R:

$$REI = \int C_i dt / \int C_R dt \quad | \quad C_i = local$$

The air distribution system may be evaluated for acoustic acceptability by measuring the sound pressure levels at the eight octave bands defined in the Room Criteria Curves, and comparing the results with the recommended Room Criteria for the functional area.

The air distribution system may be evaluated for energy efficiency by determining the energy requirements for each of the occupied spaces served by the system, ΣER_i, and the energy consumed by the system, EC. The ratio (i.e., energy efficiency) may then be compared to the evaluation criterion of 70%:

1. The energy requirements should be determined by establishing energy balances for each of the occupied spaces, shown as Module 1 in Figure 14.
2. The energy consumption of the system may be obtained by direct measurement of the fuel and electrical usage, or by establishing energy balances for the other relevant modules shown in Figure 14.
3. The air leakage may be determined by subtracting the measured room air supply rates from the value obtained from the pitot traverse taken in the supply air duct of the air handling system, upstream from the first branch take-off.
4. The Air Transport Factors may be determined by summing the products of the measured room air supply flow rates and the difference in room air supply and return air temperatures, and by dividing this sum by the total power required by the supply and return fans:

$$ATP = (1.1\ V_{s,i}\ \Delta T)/\{3414\ (kW_{supply} + kW_{return})\}\ fans$$

5. The fan efficiency ratio may be determined by dividing the fan efficiency at operating conditions by the maximum fan efficiency obtained from catalog or manufacturer's data. The fan efficiency at operating conditions may be determined from the measured flow rate, V_i (cfm), the external static pressure of the fan, ΔP (inches H_2O), and the fan motor power, HP:

$$E = (V_{s,i} \Delta P)/(6356\ HP)$$

Conclusion

Although seldom accomplished in practice, it is feasible to evaluate the performance of an HVAC system objectively. With new awareness developing about the impact of energy efficient design and operation on indoor air quality, the time when we should begin employing performance evaluations of these systems may have now arrived.

References

1. Committee on Indoor Pollutants: *Indoor Pollutants.* Report by the National Research Council. National Academy Press, Washington, DC (1981).
2. Woods, J.E. and E.A.B. Maldonado: *Final Report. Development of Energy Management Program for Buildings in Iowa-Fourth Year*, Vol. 1, *Development of a Field Method for Assessing Indoor Air Quality in Single Family Residences.* Iowa Energy Policy Council contract 82-4000-03. Iowa State University, ISU-ERI-Ames 82469 (May 1982).

3. *ASHRAE Standard 55-1981. Thermal Environmental Conditions for Human Occupancy.* American Society of Heating, Refrigerating, and Air-Conditioning Engineers, Atlanta (1981).

4. **Committee on Industrial Ventilation:** *Industrial Ventilation — A Manual of Recommended Practice,* 17th ed. American Conference of Governmental Industrial Hygienists, Lansing, MI (1982).

5. *ASHRAE Standard 62-1981. Ventilation for Acceptable Indoor Air Quality.* American Society of Heating, Refrigerating, and Air-Conditioning Engineers, Atlanta (1981).

6. **Berglund, B., U. Berglund and T. Lindvall:** Characterization of Indoor Air Quality and "Sick Buildings." *ASHRAE Trans 90(1)* (In Press) (1984).

7. **WHO Working Group:** *Health Aspects Related to Indoor Air Quality.* EURO Reports and Studies 21, pp. 1-32. World Health Organization, Regional Office for Europe, Copenhagen (1979).

8. **National Technical Information Service:** *Building Control Law and Dust Collectors.* (In Japanese; English Abstract) APTIC No. 63252 (1974).

9. **OSHA:** Occupational Safety and Health Standards. Subpart Z — Toxic and Hazardous Substances. *Code of Federal Regulations,* Vol. 29, Part 1910. U.S. Government Printing Office, Washington, DC (1982).

10. **Schaefer, K.E., Ed:** Preventive Aspects of Submarine Medicine. *Undersea Biomed. Res. 6 (Suppliment):*S-1-S-246 (1979).

11. **Parker, J.F. and V.R. West:** *Bioastronautics Data Book,* 2nd ed. Pub. No. NASA SP-3006. NASA Washington, DC (1973).

12. **Janssen, J.E., T.J. Hill, J.E. Woods and E.A.B. Maldonado:** Ventilation for Control of Indoor Air Quality: A Case Study. *Env. Intl.* 8:487-496 (1982).

13. **Maldonado, E.A.B. and J.E. Woods:** A Method to Select Locations for Indoor Air Quality Sampling. *Building and Environ.* 18(4):171-180 (1983).

14. **Nevins, R.G. and P.L. Miller:** Analysis, Evaluation and Comparison of Room Air Distribution Performance — A Summary. *ASHRAE Trans.* 78(2):235-242 (1972).

15. **Nevins, R.G.:** *Air Diffusion Dynamics.* Business News Publishing Company, Birmingham, NJ (1976).

16. *ASHRAE Handbook and Product Directory: 1980 Systems Volume.* Chapter 35, *Sound and Vibration Control.* American Society of Heating, Refrigerating and Air-Conditioning Engineers, New York (1980).

17. *Manual of Procedures for Authorized Class A Energy Auditors in Iowa.* Prepared for Iowa Energy Policy Council by Engineering Research Institute, Iowa State University, ISU-ERI-Ames-79076 (1979).

Comfort and discomfort in office environmental problems

JOHN A. CARLTON-FOSS, Ph.D, S.M.
President, Human-Technical Systems, Inc.*

Orientation

An important part of successful design and construction of buildings is that they be both safe and comfortable to occupants, both when completed and also many years after completion. Thus it is important to be able to determine what conditions and designs are safe and comfortable. Because the evaluation of people's comfort and health, the built environment, and the solution of building-associated problems are necessarily interdisciplinary and complicated, this paper can and should be understood from a number of different perspectives. One of the important perspectives is that of how to diagnose a building. A little more than a year ago I was walking down a hallway with a colleague, having discovered that we both embraced the idea of "people as sensors" for measuring the quality of the environment. However, he was more inclined to take people's evaluation at face value, while I believed that it was crucial to "calibrate" the sensors before drawing any inferences. People are more complicated than sensors, and so one might alternatively ask, "Can people's reports of comfort, discomfort, and symptoms be used as a means of evaluating the environment in a building?" The simple answer is "yes, but" one must be very smart and very careful. A more sophisticated answer would include a number of criteria such as the following: one should a) understand and be able to apply the basics relevant to the comfort and health of large samples of people; b) whenever possible use questions and procedures already validated to gather data and analyze it; c) pay attention to individuals and their differences; d) set up cross checks with methods (e.g., employee performance) based on other criteria than "comfort;" and e) keep costs reasonable in terms of dollars and requirements on employee time. The rest of this paper will present a third level of sophistication in evaluating building environments from the point of view of the people who occupy them.

Introduction

The effect of the indoor environment on people's comfort and health depends on a large number of factors including thermal factors and individual and organizational factors. While some building problems are characterized by a single factor, usually we find two or more factors interacting with each other. Thus when a problem exists, many factors must often be considered, including the thermal environment, office and mechanical equipment and its performance, what that equipment requires of people, and what people's feelings are about it. There are also management, job design, and frankly political dimensions to problems in the workplace.[1,2] When complaints arise, the presented problem often is not the actual problem. We have, for example, been called into look at complaints about lighting in a VDT room, only to find that the most significant problem was temperature.

I will be focusing on the air itself, its thermal characteristics, on the buildings through which it flows, on its movement as it affects people's comfort, and on the ties between these and non-thermal attitudes of environmental quality. Other factors will enter the picture as they interact with thermal factors. In part, thermal discomfort can make people more edgy, fostering less lattitude for adaptation to other problems of the office; these in turn can make people edgy, decreasing their flexibility to thermal environments that do not suit them.

The quality of indoor air is of increasing concern to many of us responsible for the indoor environment, for the health and productivity of people, and for the success of organizations. In my experience significant air pollution or thermal discomfort can be found in almost every building if a person

---

* Lincoln Center and Weston, MA 01773; 617/891-8050

looks a little, but often the causes of such problems are elusive. The problems have often been disregarded by all but the the chief building manager or the industrial health official, who are responsible for preventing or resolving them, too often with inadequate financial and institutional support.

It is difficult to differentiate between complaints stemming from poor quality office environments and "gripes." In my experience there have always been causes and remedies for complaints, but identifying them often requires very hard work. This includes physical factors, but one must also usually understand the people and organizations involved. Psychological factors almost inevitably contribute as antecedents, concomitants, and consequents of comfort and health problems and solutions. In the extreme, complaints can be demonstrated to have no physical cause, and may then properly be labelled "psychogenic." As I discuss in a later section, I am extremely wary of this diagnosis.

Of great concern in the present highly competitive, uncertain economy are the performance deficits which may result from people's exposures to polluted or thermally stressful indoor air. These physical factors may mix with organizational ones to yield severe problems. Such issues are very new and require major work. I will begin with issues of thermal comfort which have been well studied. Following this I will discuss individual differences, the role of personality in comfort, and the relation of "comfort" to other measures of environmental quality. It should be noted, though, that I will focus on the state-of-the-art of building-associated comfort problems, not review in detail the many established areas of research that have led to it.

Thermal comfort

Discomfort due to poor thermal conditions and indoor air pollution has attracted considerable attention. In the late 1970s[3] the number of complaints became quite significant. It appears that people had become alerted to the problem of chemical and other pollutants in their environments,[4] and also that indoor air quality problems had been exacerbated by a common belief that one could successfully minimize energy costs by (sometimes dramatic) reductions in ventilation. In the late 1970s the Emergency Building Temperature Restrictions (EBTR) came into effect. Formulated to reduce energy use while avoiding significant short-term health problems, these regulations did nothing to help people adapt, little to maintain occupants' comfort or productivity, and a lot to maximize complaints about and discontent with the environment. Recent work suggests that being in an office environment that is too warm and humid may also lead to complaints about "stuffiness," "stale air," and "poor indoor air quality."[2] Thus it is very important to address thermal issues in any approach to problems in buildings.[A]

Human response to thermal environments has traditionally been studied as a function of eight variables: 1) dry-bulb temperature, 2) mean radiant temperature, 3) water vapor pressure, which is used by thermal comfort researchers in place of relative humidity, 4) air velocity, 5) clothing level, 6) activity level, 7) time, and, occasionally, 8) heat storage in the body tissues. In principle, these are all objective, measurable statistics. Models of thermal acceptability and comfort have used them as parameters and have offered an impressive, albeit approximate, description of environmental conditions for human comfort and discomfort. In light of those apparent successes, care must be taken also to consider individual differences among people. There are limitations in the comfort equation which serves as the basis for the new ISO standard, and new findings are emerging in the field. If it were not for these limitations, we would be able to calculate environmental conditions to satisfy everyone, and would only have to worry about what clothing they wore and how active they were.

Fanger's[7] physiologically-based, environmental engineering theory of thermal comfort models the human body in much the same way that an engineer models a building. Both the building and the body have an outer surface through which heat is gained or lost. The amount of heat transfer for given internal and environmental conditions depends upon the total surface area and insulation-value of the surface. Both building and body surfaces are permeable to heat and vapor transfer,

---

[A]Increased use of automated office equipment and an aging of the workforce also may require more favorable environments than have traditionaly been thought necessary in offices.[5,6]

```
Vote No. _____          Test No. _____
Number _____  Sex _____  Name _____
         Circle the number beside the adjective
              that describes how you feel.
           9    Very Hot
           8    Hot
           7    Warm
           6    Slightly Warm
           5    Neutral
           4    Slightly Cool
           3    Cool
           2    Cold
           1    Very Cold
```

Figure 1 — Examples of a one-dimensional thermal comfort ballot.

the magnitude and direction of which depend on internal and environmental variables. For example, the rate of conductive heat flow varies linearly with the temperature difference between the interior and the exterior. Additional factors are mean sol air temperature (for buildings) and mean radiant temperature (for people), the velocity of air in the environment, the water vapor pressure in the environment, and the amount of heat gained internally. Fanger and ASHRAE *Fundamentals* (Chapter 8) present explicit equations for internal heat generation, heat loss by skin diffusion, heat loss by evaporation of sweat, latent respiration heat loss, dry respiration heat loss, heat conduction through clothing, heat loss by radiation, heat loss by convection, and heat balance. The net heat gain/loss must be null for the body to be in thermodynamic equilibrium. On the average, the closer a body is to thermal equilibrium, the more comfortable the person is presumed to be. However, care must be taken in using Fanger's equation, for it does not adequately represent evaporative heat lost,[8] does not apply to individuals,[8] underestimates the effects of relative humidity on comfort,[9] and is of limited accuracy in representing thermal *dis*comfort.[10] In spite of these shortcomings and its simple assumptions about the nature of people, Fanger's theory does a good job of explaining the average thermal responses of large populations as long as they are comfortable or nearly so.

Measuring comfort and discomfort

Judgements of the comfort and/or acceptability of thermal environments have been scored on ballots with a one-dimensional scale, or a seven-dimensional semantic differential scale. Subjects are given a sheet of paper with the scale on it, and asked to indicate their evaluation of the environment. An environment is regarded as "acceptable" if at least 80% indicate that it is "comfortable." This approach has also been used to set standards for current work on comfort-based acceptance of odors, chairs, and of indoor air quality in general. Figure 1 is a sample of the traditional one-dimensional ballot[11] and Figure 2 is the seven-dimensional ballot developed by Rohles.[13,14]

Subjects' responses to each of the scales are multiplied by a weighting factor, the products summed, and the sum divided by a normalization

```
                    Thermal Comfort Ballot
    According to the instructions, place a check between each pair of adjectives
    at the location that best describes how you feel:

              comfortable ___:___:___:___:___:___:___:___ uncomfortable
           bad temperature ___:___:___:___:___:___:___:___ good temperature
                 pleasant ___:___:___:___:___:___:___:___ unpleasant
                     cool ___:___:___:___:___:___:___:___ warm
             unacceptable ___:___:___:___:___:___:___:___ acceptable
    uncomfortable
       temperature ___:___:___:___:___:___:___:___ comfortable
                                                     temperature
                satisfied ___:___:___:___:___:___:___:___ dissatisfied
```

Figure 2 — Example of Rohles' seven-dimensional thermal comfort ballot.

RH (%)	\multicolumn{20}{c}{Dry Bulb Temperature (°F)}
	60	62	64	66	68	70	72	74	76	78	80	82	84	86	88	90	92	94	96	98
85	0	0	2	0	4	1	2	5	7	3	6	2	1	1	1	0	0	0	0	0
75	0	0	0	0	0	2	7	1	8	8	2	2	2	2	1	0	0	0	0	0
65	0	0	0	0	0	1	2	8	4	6	5	2	0	3	2	1	0	0	0	0
55	0	0	0	3	1	1	3	5	8	4	6	4	2	3	0	0	0	0	0	0
45	0	0	1	0	3	2	2	1	8	7	9	6	5	2	3	1	0	1	1	0
35	0	0	0	0	0	1	1	1	8	6	7	6	4	5	1	0	0	0	0	0
25	0	2	0	0	0	1	2	3	2	6	6	4	6	4	4	3	2	0	0	2
15	0	0	0	0	0	0	3	1	2	6	7	6	2	7	1	0	4	0	0	1
TOTAL	0	2	3	3	8	9	22	25	47	46	48	32	22	27	13	5	6	1	1	3

Figure 3 — Number of subjects out of 10 subjects per cell who recorded votes of "comfortable" after a exposure of 3 hours. Note that the sample size is 10 per cell, so that the individual differences of subjects assigned to each cell are likely to be noticeable, and significant random variations can occur. To tie such results together to generate a comfort zone of the heat balance equation, it is necessary to perform statistical averaging. Such averaging is necessary and appropriate for establishing standards, though industrial hygienists concerned with problems in the field must also place emphasis on individual people and situations.

constant to yield the percent comfort rating with (100% being perfectly comfortable). This approach is important for developing the next generation of evaluations of pollutant-modified indoor air quality because it acknowledges the varying words (i.e., concepts) people have for describing different aspects of their environments.

We have found, however, that even such an improved scale is not enough to capture an accurate report of people's comfort or discomfort. In addition, information must be gathered on the adaptation level[14] of a subject, that is, on what environmental conditions subjects prefer and what words they use to describe those conditions. In thermal comfort research the issue is whether the subject regards "warm" as "good," or "cool" as "good." Exactly the same issue comes up with respect to indoor air contaminants, some of which are noxious, some others even being fragrant. The importance of such information became obvious when a subject circled "8" (= hot) on a one-dimensional ballot and then wrote on the ballot: "*I like it that way.*" Not surprisingly then, personality and the level of thermal sensations to which people have adapted became key dimensions in the research reported below on individual differences in environmental response.

The "comfort zone" and asymmetric thermal environments

The procedures outlined above have been used to determine the frequencies with which people feel comfortable in various temperature conditions. Figure 3 makes clear that there is a great variety of uniform-temperature regimes in which one or another person may feel comfortable, or uncomfortable. So, when addressing a complaint in a building, it may not be possible simply to refer to the thermal comfort zone described in ANSI/ASHRAE *Standard 55-1981*,[15] or in *Industrial Ventilation*.[16] Even so, it is important to note that one can derive this comfort zone from the experimental data, and that if one is setting uniform environmental conditions (a different purpose than resolving a complaint or a health problem), they should probably be somewhere in this "comfort zone" in which about 80% of people can be expected to express satisfaction with thermal conditions.

There are additional extremely important components of the thermal regime of offices. These include asymmetries in the radiant temperature. The classical case is the person who sits next to an outside window and complains of being too cold during winter, even though the thermostat keeps the dry bulb air temperature in the comfort zone. This is a particularly important point to note, because it is not covered in *Industrial Ventilation*. Unilateral heating or cooling, the more general term, can be caused by:

1. An asymmetric radiant field. To a person walking outdoors, the sun creates a radiant temperature of 30°-40°C.[17] People sitting by sun-facing windows in summer thus are exposed to much warmer temperatures than those in the rest of a building. In addition, the source of the heat is localized on one side, so it might be expected to produce a different effect than if the radiant temperature were the same on all directions.

2. Local high or low air velocity. If the air velocity is high, people may feel cold or may complain of draughtiness, whereas if air velocity is low, they may complain of "stale air" or of "stuffiness."

3. Warm or cold walls and floors. Outer building walls are almost invariably cooler or

warmer than the interior and its ceiling and walls. This results in cooler or warmer perceived temperatures for occupants located in interior zones. Because the interior and exterior zones may have different heating and cooling systems and be made of materials with different thermal properties, the thermal environment in interior zones may, for example, take longer than in the exterior zones to warm up on Monday morning. Thus, the air temperature may be in the comfort zone, and people may be satisfied with the thermal environment for most of the week, but there will still be complaints on Monday morning after the heating system has been set back for the entire weekend and building materials must be heated up.

Radiant temperature asymmetries warrant further treatment. For short-term exposures to young, healthy people, these asymmetries can be quite extreme and still yield comfortable conditions — so long as the net heat flow is balanced between the entire body and the environment. An extreme example, utterly inapplicable to office building occupants, was produced in an experiment which addressed the possible effects of spacecraft environments on astronauts during reentry,[18] the researchers exposed subjects' anterior sides to radiant temperatures of 82°-93°C, and the posterior sides simultaneously to radiant temperatures of -7°C. The air temperature was held at 21°-24°C. In spite of such extreme conditions, the subjects reported overall comfort.

It is in long-term exposures of older or younger, presumably less healthy people to less extreme radiant temperature asymmetries that there may be a significant effect to beware of. Boje, Nielson, and Oleson[19] exposed 32 young and old subjects to unilateral cooling by panels at 10°C for 15 days at six hours per day. Fanger summarized the results as showing that

> "... the unilateral cooling caused no discomfort ... but for a few of the subjects (in the course of the experimental period) [a palpatory examination found] *a thickening of the cutaneous and subcutaneous tissues and an increased tension and soreness of the muscles on the side of the body exposed to the cold panel.*"

Of course, many office workers near exterior windows are exposed 5 to 7 days per week for months or years to similar conditions. Thus, one can easily envision in office workers effects that will be similar, or even more dramatic.

One must therefore address the complaints of such people with great sympathy and analytic rigor. As an example, consider the bookkeeper sitting near a NE-facing window in a 70,000 square foot office building. (Her location is denoted by the "B" in Figure 4.) Her complaint was "draftiness." She sat with her back to the northeast. Measurement with an Alnor Series 6000P Velometer revealed air speeds significantly less than 50 feet per minute. Simple reference to the comfort zone would indicate that she "should" be comfortable. Yet she was not at all comfortable. Was she specially sensitive to drafts? In an informal conversation, she said that she was. Yet in this case, such sensitivity might not be entirely a "personal problem" or a matter of "personal sensitivity." Her exposure, at age 50, to long-term radient asymmetric fields would make her more sensitive to drafts, and also might put her on edge, so that she would be bothered more by the drafts she noticed.

Cleary, then, careful attention must be given to physical factors as well as the ages of people in the workplace. It appears that exposure to certain physical conditions can even cause certain individual differences among actual occupants of buildings. However, there are other sources of individual differences which are of at least equal importance.

Individual differences

To really understand individual differences and psychological factors in thermal comfort,[8,21] we need a more sophisticated theory. We have found that, for example, low rather than high anxiety and the degree to which people are the origins of their relation with their environments are correlated with a broader "bandwidth" of environmental conditions that people will find acceptable. (*Note:* I did not say "prefer" at the end of this sentence.) In other words, people who are "origins," will tend to be more adaptable than pawns.

There were three independently measured variables in a field study testing this theory. "Origin-pawn" measured the extent to which a respondent's self-concept (i.e., in this case, the manner of structuring events in life) placed control within or out in the environment. Frederick Rohles' comfort index (RCI) measured respondents' percent satisfaction with the thermal environment. "Corrected temperature" was a measure of the temperature in an environment after variations in relative humidity, clothing and activity level were normalized to a preselected standard. "Corrected temperature" differs from "effective temperature" only in being normalized to 40% relative humidity. These three measured variables are plotted on the graph in Figure 5, each point corresponding to one respondent. The vertical axis is origin-pawn (origins corresponding to plus numbers toward the top of the scale and pawns to minus numbers at the bottom of the scale); the horizontal axis is corrected temperature. Contours are drawn around groupings of thermal comfort scores to assist in visualizing the shapes of the scatter plots for those reporting relative thermal comfort (the RCI greater than 50%) and for those reporting relative thermal discomfort (the RCI less than 50%). Of the 41 people in the sample, 29 had a comfort index score greater than 50%. Two of the people provided data that did not fall within either contour; both of them were tested at home during the evening after work, while all the rest were tested in public places.

Correlations between personality and comfort were significant at the 0.02 level and origin-pawn accounted for 15% to 20% of the variance in comfort of people who were acceptably comfortable, and for 30% to 37% of the comfort level of those who were uncomfortable.

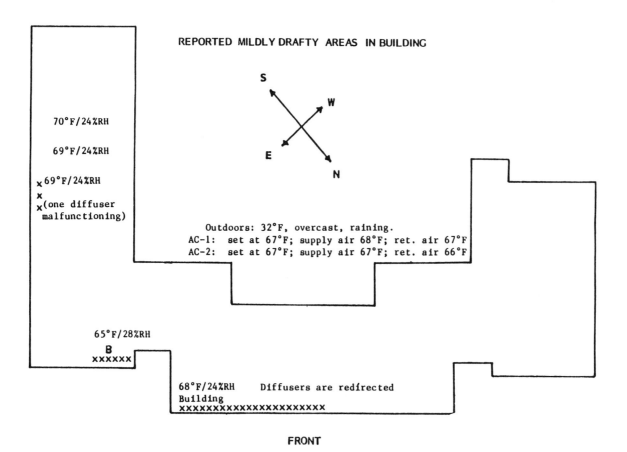

Figure 4 — Distribution of temperatures and selected environmental problems on first floor of a 70,000 square foot building. The person located at "B" complained of draftiness, but was also subject to a significant radiant temperature asymmetry. Note the temperature gradient of about 5°F from the core of the building (where the thermostats were located) to the periphery (where most of the people were located.

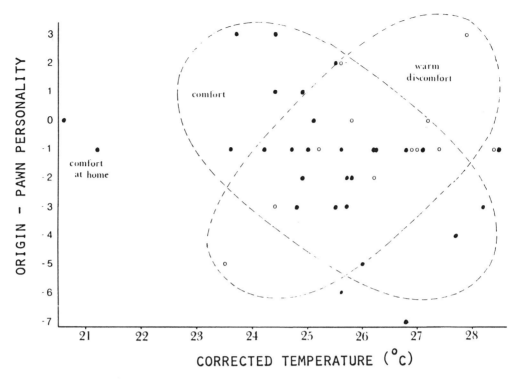

Figure 5 — Thermal comfort, origin-pawn, and corrected temperature. The scatter plot above represents data points for the research. The vertical position of a point represents the origin-pawn score for a subject (origins at the top or (+)-end of the scale, pawns at the bottom or (-)-end of the scale). The horizontal position of a point represents the corrected temperature for a subject. The solid dots represent subjects who reported thermal comfort (RCI) greater than 50%, while the open circles represent subjects who reported thermal comfort no greater than 50%. The oval labeled "comfort" highlights the approximate envelope of the data points for people reporting thermal comfort in organizational settings. The oval labeled "warm discomfort" highlights the approximate envelope of the data points for people reporting too warm in organizational settings. The points to the left of both ovals are labeled "comfort at home," signifying that they reported thermal comfort greater than 50% while at home. This suggests that such variables as anxiety and "feeling at home" are significant factors in comfort and discomfort.

There are at least two crucial points underlying this. One, if you use people as "sensors," you must "calibrate" them. You must identify their likely biases so you can make corrections and know what their evaluations really imply about the environment. Two, if no attention is given to individual differences and their source, the data indicates only that there is a scatter of points about a central temperature. There are no significant correlations, and we have no way to distinguish individuals who are thermally uncomfortable from those who are thermally comfortable. However, when attention is given to subjects' thermal sensation adaptation level and to their personalities, we do get a temperature dependence, and we can begin to distinguish those who are uncomfortable from those who are comfortable.[B] Theoretical work in this area[20] indicates several addition psychological variables which could be used to develop even better models for such individual differences.

Psychodynamic and motivational theory: toward criteria for "mass hysteria"

One of the major problems of work with complaints of "stuffiness" or "poor indoor air quality" (aside from the fact that we have no operational definition of "stuffiness" and have only anecdotal information on what it really is) is the tendency to discount the psychological domain and credit only measurable physico-chemical sources of illness in buildings. So when nothing specific can be

---

[B]The author expresses his appreciation to George Rappolt for pointing this out.

identified as "the cause" of complaints, problems are often labelled "psychogenic." This amounts to a logical doublethink: first people are eliminated from consideration, then blamed when the technical analysis does not yield a cause, when in fact these very people could have helped in identifying the cause(s). Of course, people can provide significant causes themselves (e.g., attention must be given to eating habits which may increase susceptibility to disease). However, there is a tendency to treat as "psychological" that which cannot be measured or understood. It can also go the other way: I know of a sociologist who places overdue emphasis on the people, and largely ignores technical factors. This emphasis on disciplines too often gives psychologists a bad name among technical people, and technical people a bad name among psychologists. In fact, we must all work together.

When no specific physical causes have been identified, we must not permit facile conclusions that there is "hysterical contagion" in a building. Instead, we must look to see if the pattern of those with symptoms correlates with the pattern of susceptible personality types, and if the pattern of symptom-free people correlates with that of people of low susceptibility to such contagion. It is also necessary to find out if those with complaints and pathologies are not simply more aware, sensitive, exposed, and/or able to make themselves heard than the others.

It is possible to list specific criteria for susceptibility to hysterical contagion. These criteria can help us distinguish between those highly susceptible to psychogenic disease, and those relatively immune to it. Criteria include:

1. Demanding dependency rather than mutuality.
2. Vague, impressionistic perception of the environment and the self.
3. Use of repression as a defense mechanism.
4. Poor logic, or spotty reliance on logic in reaching contact with reality; using hunches as if they were reality.
5. Incapacity for persistent or intense concentration.[21,22]

Items 1, 2, 4, and 5 are similar to the "pawn" personality, while their opposites approximate the "origin" personality.[8,20] These characteristics are quantitatively and objectively measurable, and can be used to inform field studies.

As a preliminary test of this theory we studied a 50,000 sq ft problem building in the greater Boston area. We were brought in because there were extensive complaints of "stuffiness" in the computer and personnel departments. We performed detailed instrumental measurements of environmental conditions, and distributed questionnaires to the occupants. We used exactly the same methods we used in our earlier work on thermal comfort.[8,20,23] Air velocities were well under 10 feet/minute, in part because of high open office partitions and in part because of attempts to reduce ventilation to save energy. Temperatures varied from 70.5°F in the computer room to 81°F in a clerical area of the personnel department (see Figure 6). However, the occupants' evaluations of their environment did not correlate with the measurements. Instead, they correlated with the personality of the occupants. Temperature and other environmental variables accounted for less than 1% of the variance in occupants' evaluation of their environments, while occupants' scores along the origin-pawn dimension accounted for 41% of the variance. Origins were significantly less satisfied with the indoor atmosphere than were pawns. Since the environment was in fact "stuffy," we concluded that the origins had been perceiving their enviroment accurately, and were doing something more constructive than merely "griping" or manifesting hysterical symptoms in relation to their indoor environment.

Stress and reactions to the environment

The psychodynamic theory of stress, disease, and health provides an underlying model for responses to indoor atmospheres. (Readers who are invclved with engineering and not interested in clinical issues may wish to skip this and go on go the next section.) Selye[24] posits that stressors internal and external to the organism impinge upon it, creating ever higher levels of stress. In a broad sense, any event requiring adaptation or assimilation is stressful, though it should be kept in mind that reasonable total stress levels may be desirable. Low temperature is one example of an environmental stressor.[25] Poor air quality is another, as are many other factors related to the

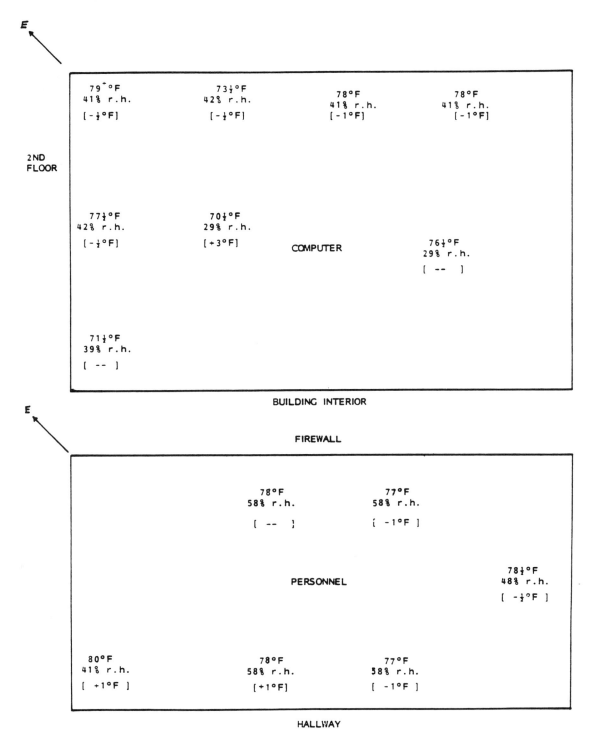

Figure 6 — Distribution of dry bulb temperature and relative humidity in a 50,000 square foot office area in which there were complaints of "stale air."

senses as well as to the mind and body. Stress is dissipated by the organism, but might not be dissipated as rapidly as it is accumulated. Once the stress reaches a certain threshold, the organism begins to malfunction (Figure 7). This "disease" may take various forms, e.g., undesirable behavior, performance deficits, medical illness, death. Common to all such "diseases" is the general stress reaction, the "general adaptation syndrome." Part of this syndrome, stimulation of the sympathetic nervous system resulting in vasoconstriction in the extremities and piloerection,[26] is readily interpreted as cold discomfort. In fact it can lead to physiological cooling of the extremities, and to other symptoms of arousal, all of which individuals will be inclined to interpret in one way or another. More generally, Schachter[27] has reported on work indicating that people may interpret internal sensations associated with arousal in any one of a variety of ways, depending on their inclinations and the demand characteristics of their environment. Frank[28] states this succinctly when he writes that Schachter "showed that persons who are physiologically aroused are impelled to search for a reason and that the emotion they actually experience seems to be a fusion of the state of arousal with the person's attribution of its source." This suggests that there will be individual differences in peoples' responses to the environment: some people will be more inclined to interpret their internal sensations (of whatever cause) as discomfort or illness than will others, and some people will be more likely to attribute that discomfort or illness to environmental factors than will others (Figure 6). Stress and peoples' inclinations to interpret its signs one way rather than another, can be understood more systematically and completely in psychodynamic terms.

Individuals under stress develop adaptive and maladaptive functioning in stages[29] (see Figure 8). The details of the function and dysfunction depend upon individuals' ego organization.[30,31] Many predictive hypotheses can be derived from this psychodynamic theory, though it is also extremely important to emphasize that people untrained in psychology must not be permitted to give facile diagnoses of "personal problems" as a means of avoiding mitigation of real air quality problems. If people use somatization as a defense at a given level of stress, they might tend to become physically sick in response to sufficient stress. (Note that this stress is the accumulated stress of internal and external origin from activities in all sites that a person has recently frequented.) If they use passive-aggressive defenses, they might tend to complain behind the back of the person who controls the environment. If they use displacement, they might think the office is cold or polluted when actually they are angry with their supervisor. If they use denial, they may invest extreme energies in maintaining their exposure to substances that will harm them, meanwhile denying (without awareness, and sometimes with apparent aplomb) that these substances are really

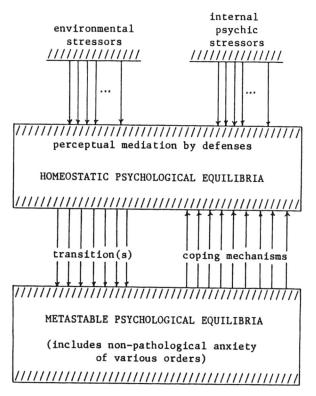

Figure 7 — Psychodynamic model of stress and discomfort. In this model stressors induct transitions of the psyche from its initial states of homeostatic equilibrium, to one or more metastable, less desirable equilibrium states. The organism uses coping mechanisms to resist the transition process away from homeostasis, and also to make transitions back to homeostatic region. Each organism has its own preferred mix of homeostatic equilibria, metastable equilibria, transition making, and use of coping mechanisms. Each organism is also subject to its own mix of internal environmental stressors. Note that this model does not include any of the psychotic states, whether stable or metastable or unstable, which could conceivably result from stress.

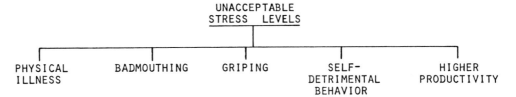

Figure 8 — Some alternative channels for release of stress tension in the workplace.

harmful to themselves or to anyone else. If they use sublimation (i.e., this may mean that they are highly motivated), they might tend to work harder and to be more productive until pollutants yield significant detrimental effects. If they are accustomed to taking effective, constructive action to cope with and find remedies for problems, they might tend to do the same in the face of environmental stressors.

The impact of the stress due to mild environmental discomfort may be small in most cases, but even so it might produce perhaps a 5% shift toward the use of less mature defensive styles (Figure 9), and this shift might be crucial. If less mature defensive styles are used even slightly more frequently in situations in which more mature styles are required, the impact on delicately balanced personal and social systems could be enormous. It should also be recognized that there will be cases in which apparently mild stress increments will lead to major regression, or (particularly for people under medication) even to death.[32-36]

Building associated illness: thermal causes?

How then can we measure and diagnose conditions which cause problems? How can we formulate and carry out laboratory research that is relevant (i.e., is valid and has high external reliability)? Considering the technical and psychological factors involved, the work is not at all straightforward. Even highly skilled specialists will necessarily find it a considerable challenge. Clearly, we must be very sophisticated in how we solicit peoples' responses to their environments. Most of these people have no firm idea what causes their responses to their environments. They experience sensations which they interpret on the basis of

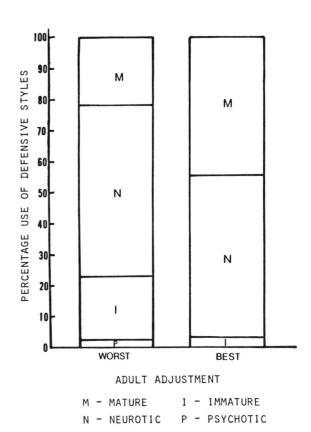

Figure 9 — Composite pattern of distribution of defenses for the psychologically most healthy versus the psychologically least healthy people in a study of the psychological health and career success of Harvard graduates. Note that there is only a 25% difference in the frequency with which these two groups exhibit healthy, mature behavior. The greatest differences occur in the extent of use of suppression (consciously living with but setting aside uncomfortable, presently unresolvable issues in one's life) and anticipation (anticipating difficult life events and preparing for them). Both of these are likely to be used less frequently when a person's "slack is used up" adapting to difficult environmental conditions. In their place are the less mature behaviors and patterns of thought, which will tend to disrupt the effectiveness of organizational systems.

personal beliefs, personality, or interests. They are influenced by situational cues including the interpersonal ambience in the workplace or laboratory. One "building associated epidemic" seems to have been triggered by a new odor which appeared in the office after a long period of smoldering discontent about the indoor environment.[37] Experimenter-subject effects (e.g., positive and negative Hawthorne effects and placebo effects) are well known, but seem to have been addressed primarily by Rohles[38,39] and Carlton-Foss.[2]

If investigators are to determine what remedies might actually be useful, they must determine the real problems in a building. Complaints of "poor air quality," "lack of air motion," "stuffiness," or other subjectively evaluated conditions must be investigated on several fronts: airborne contamination, thermal conditions, engineering and architectural issues, and personal and organizational dimensions. Some investigators have put considerable effort into measuring the composition of the air with complex equipment, while others have not even recorded temperatures. Some investigators have given out extensive but unfocused and unreliable sociological questionnaires to gather the opinions of building occupants, acquiring little useful personal or organizational data,[2] while other, more sensitive investigators have serendipitously discovered hidden causes of air quality complaints such as improper construction or operations, a simmering labor-management problem, or disgruntlement about uncomfortable temperatures.[41,42]

But people in many normal buildings probably *should* be complaining about "stuffiness."[42] Interior zones are cooled all year. In many variable volume (VAV) system operations, thermostats in the interior of large buildings are satisfied by heat migration from the interior to the perimeter. Thus, VAV in interior zones operates at minimum volume for most of the time and ventilation is inadequate.

In addition, with such low air flow rates, it is unlikely that enough moisture is removed from interior spaces. Thermostats typically used in buildings sense only the temperature and not the high humidity levels that are often created by occupants, plants, equipment, and weather. At these humidity levels, people not only require lower temperatures for comfort (Figure 10), but also are apt to be more aware of odor problems.

Air movement is also important. Of course, partitions and acoustical panels can obstruct air motion, but air diffusion technologies are also an issue. Air diffusion systems at volumes of 0.6 cfm/ft^2 (cubic feet per minute of ventilation air per square foot of floor space) and velocities of 20 feet or less per minute can maintain temperatures within ± 0.3°F throughout a space. However, no known air diffusion system can maintain uniform air motion between 45-80 fpm, the air speed required to meet the requirements of ASHRAE Standard 55-1981 for comfort at 78°F and 1.0 CLO.[42] So air motion is likely to be a source of problems in buildings left warm in summer or winter.

Bruce Findlay[43] reports that most air quality complaints in his Canadian government buildings have been resolved simply by adjusting the thermostat by a few degrees. He found, for example,

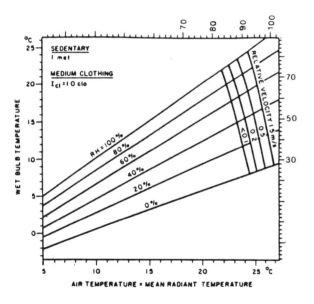

Figure 10 — This graph based on the Fanger comfort equation illustrates the relationship between relative humidity, temperature and comfort. Lines of constant comfort have been drawn for selected values of air velocity (AV) and relative humidity (RH). As RH increases for constant AV, the comfort temperature decreases. As AV increases for constant RH, the comfort temperature increases. Examination of this chart reveals one of the limits to ventilation for high air quality: if large quantities of ventilation air flow through the room, the AV may become too high for comfort in local areas of an occupied space. Generally, occupants should be assumed sedentary in the interior zone of a building where there is low air motion.

that the costs are not always great for changes from 77°F to 72°F during the summer. He reports only a $1000 increase in summer cooling costs for a 22,000 m² modern office building (with an economizer cycle) in Ontario, and much reduction in complaints. This offers anecdotal corroborating evidence for the hot air hypothesis. Unfortunately, there is no research yet begun on the subject.

Comfort and other measures of indoor air quality

In general, building owners and managers have tended to use complaints as their index of the acceptability of environmental quality. This is risky because complaints are often a biased, discontinuous measure. They may also be registered by special people, and they are not frequently tied in with office politics. Thus, there may be serious medical problems by the time a real problem is acknowledged.[2,44] Let me give an example. Recently we began a pollution abatement project at the request of the president of a small organization. He was sick from indoor air pollutants, and his doctors (knowing nothing about industrial hygiene) had told him that his only choice was to quit the business. Yet no one else in the organization was affected, and the men on the job floor had managed more or less to convince him that the problem must therefore be "psychological." So the first part of our work included convincing him that this was most likely not the case, but that some standard engineering could be used to address the problem. Unfortunately, by this time our client had already developed rather severe symptoms.

In contrast, ASHRAE has focused on comfort as the primary criterion for acceptable (indoor) air quality. The notion of comfort has been based on people's votes on scales similar to the thermal comfort scales. Using the comfort-vote technique, there is, for example, a threshold of acceptability at about 0.1 ppm for formaldehyde in indoor air[45] whereas 1.0 ppm is OSHA's standard for medical danger for people in the workplace. Such comfort-type scales have also been applied to research on comfort and acceptability of odors in occupied spaces.

Comfort and odor evaluations may, however, not always provide adequate measures of risk for an airborne substance. People become accustomed to odors and certain other non-thermal cues, so that for example, a smoker may be less sensitive to the effects of smoke concentrations than a nonsmoker, and longterm occupants of a room will not notice odors that will be very offensive to visitors. Further, there are individual physiological differences in people's response to odors. For example, some people might find diesel fumes objectionable because they are more sensitive to the "pungent-acid" odor component, while others may find it pleasant because they are more sensitive to the "aldehyde-aromatic" component. Further, people may differ about whether the same odor is pleasant or noxious. This highlights the need for care in using comfort as a criterion to define "good environments," and for use of corroborating measures of indoor air quality, particularly when the air pollutants in question are apt to be evaluated as pleasant rather than noxious.

There are two types of such measurement: 1) performance effects and subclinical symptomatology, and 2) medical symptomatology, morbidity and mortality. Psychological and subclinical medical symptomatology may be the first signs of, for example, carbon disulfide poisoning: victims may initially experience headache, dizziness, increasing sense of weariness, loss of strength, transient excitement, and slight delirium very like alcoholic intoxication, followed by deep depression and loss of memory, increasing indifference and apathy. This might change suddenly to acute mania or delusions of persecution with hallucinations, ending with either recovery or incurable dementia. (Note that a few of these symptoms are apt to feel "comfortable" if not actually "pleasurable" to the victims. "Glue sniffing" may provide an even better example of this.) As late as 1946, psychiatrists were admitting victims of carbon disulfide poisoning to mental hospitals with comments such as "etiology occupational" or "source unknown" and with no reference to carbon disulfide.[46] Since many of these symptoms may occur for other reasons in everyday life, meticulous differential diagnosis in research efforts and in industrial hygiene in the field is required. In addition, performance effects may be the first and most significant clues of pollution in normal living and working circumstances. For example, Weiss[47] writes that the carboxyhemoglobin levels in the blood of heavy smokers may reach levels of 5% to 7% or greater, and that this is near the threshold for impaired vigilance performance. If Weiss' view is confirmed, the implications could be considerable for heavy-smoking motorists in congested traffic. These, and other better-substantiated performance effects and subclinical symptoms often show up before acute medical ones, and in some cases can provide precise, sensitive measures of the impacts of environmental pollutants.

Studying the performance and subclinical effects of noxious chemicals could lead to serious ethical problems unless an on-site example happened to be found and exposure levels could be accurately estimated. However, a performance-oriented impact assessment methodology has been demonstrated in a study of "ergonomic" chairs.[48] Chairs, like air quality and temperature, can be considered one of many possible environmental stressors producing both general and specific effects. We hypothesized that ergonomically well-designed chairs would reduce physiological stress as well as cognitive loading on users, who would therefore be able to perform better than those who used "standard" office chairs. The people using the standard chairs would become fatigued faster because of the stress, and would be distracted by their non-ideal chairs. Thus, the study sought to measure the non-specific impacts of stress associated with use of higher quality versus lower quality chairs (analogous to the general impacts of toxic substances).

Two groups of experienced typists were seated for a full day at the same workstation, but with different chairs. They learned a standard word processing program and then used it to enter pre-typed originals into the computer. Questionnaires were administered before and after the work session to measure psychological variables. Five operators using an ergonomically sound office chair were compared with six operators (controls) using a standard office chair. Those in the ergonomic chair learned the software 20% faster, reached a 13% higher performance plateau, and maintained performance levels 40% higher during the final 1.5 hours of the 5.5-hour trial period (Figure 11). It seems that the poor chairs reduced the performance of the controls, and it seems likely that studies of the impacts of low dosages of many different environmental pollutants on people's performance of complex tasks will yield a similar pattern.

Such psychological methods can be used to measure effects that are extremely important to organizational decision makers, though they may have no obvious or immediate medical impact on people. They also provide an important check on environmental quality criteria based on the concept of "comfort."

Of course, at higher or more prolonged levels of exposure to toxic substances or probably even to low quality furniture, symptoms will be less subtle — sometimes acute. Spengler and others at Harvard[49,50] have determined that children growing up in homes with at least one adult smokers have 5% to 25% higher well-to-risk ratios than do children growing up in non-smoking homes. Tobacco smoke and other airborne contaminants have also been found to be significant contributors to fatal disease. An epidemiological study of death rates of male British doctors over 20 years, for example, showed that smokers had a death rate (standardized for age) due to chronic respiratory illness about 16 times that of non-smokers.[51]

Dose-response curves provide a more continuous measure of the effects of pollution on people. These curves represent the impacts of varying exposures to a (airborne) toxic substance or combination of such substances. Traditionally such curves have been drawn with respect to severe symptomatology and death, but performance deficits and other more subtle measures could also be used to establish relationships at the low-dose end of the curve. Such measures eventually must be correlated with more severe symtomatology in longitudinal studies. What we have now is primarily documentation of these symptoms only after they require medical attention. In the British doctor study, for example, it was found that the annual death rate for chronic bronchitis and emphysema per 100,000 men was 3 for non-smokers, 38 for those who smoked 1-14 g/day of any tobacco, 50 for those who smoked 15-24 g/day, and 88 for those who smoked 25 or more g/day. These numbers were plotted, and a curve fitted and tested using the chi-square distribution. This technique makes possible in principle determination of the level of risk (in terms of types of illness per capita or deaths per capita) for any given exposure to a toxic substance. Of course, individual differences (including those associated with psychological factors) may again complicate things. For example, Type AA people (as genetically

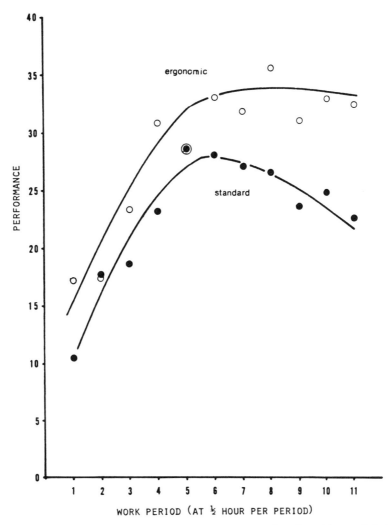

Figure 11 — Performance at a word processing task as a function of time and chair used. The first part of the graphs are learning curves with the same asymptotes and different time constants. The sameness of asymptotes implies that differences observed were not the result of differences in typing ability between samples. Once subjects had learned the word processing task but had lost their freshness, their performance leveled off. In the final time period, subjects using the misadjusted standard chair became fatiqued and their performance dropped significantly.

determined) metabolize benzo(a)pyrene into the carcinogenic material, aryl hydrocarbon hydroxylase, ten times as much as people who are not Type AA.[52,53] Thus, people who are Type AA might be exposed to a much greater health risk than non-Type AA people even though both were functioning in the same indoor environment. Behavioral differences, leading one smoker or non-smoker to be exposed to more smoke than another would result in additional differences. However, the most significant debate about dose-response curves is their extrapolation from high to low levels of exposure. To the extent that such questions can be resolved for various pollutants, dose-response curves could be useful. In this way psychological and medical measures can be used along with comfort and odor-acceptability measures to inform decisions about what constitutes an "acceptable" environment for people.

Conclusion

This paper has presented theory and work that I believe is very important for resolving issues of acceptable conditions for indoor environments. A correct approach to indoor environmental quality must, I believe, be very thorough, incorporating thermal, and psychological domains, as well as the traditional epidemiological and hygiene areas. While this increases the complexity of the task, it also provides the opportunity to resolve more of the issues of comfort, health, and performance.

References

1. Carlton-Foss, J.A.: Energy Engineering for Occupied Places. *ASHRAE J. 24*:35-39 (1982).
2. Carlton-Foss, J.A.: Tight Building Syndrome: Diagnosis and Cure. *ASHRAE J. 25*:38-41 (1983).
3. Kreiss, K.: Building-Associated Epidemics. *Indoor Air Quality*, C.S. Dudney and P.J. Walsh, Eds. CRC Press, Boca Raton, FL (1983).
4. Stolwijk, J.: Personal communication (December 1983).
5. Carlton-Foss, J.A.: *Human Factors in Automated Offices*. Unpublished manuscript distributed to selected members of ASHRAE TC 2.1 (June 1983).
6. Rohles, F.H.: Personal communication (June 1983).
7. Fanger, P.O.: *Thermal Comfort*. McGraw-Hill, New York (1970).
8. Carlton-Foss, J.A. and F.H. Rohles, Jr.: Personality Factors in Thermal Acceptability and Comfort. *ASHRAE Transactions*, Vol. 88, Part II (1982).
9. Gagge, P: Personal communication (March 1984).
10. Int-Hout, D.: Personal communication (December 1983).
11. Rohles, F.H.: Personal communication (1979).
12. Rohles, F.H.: Personal communication (1979).
13. Rohles, F.H., C. Bennett and G. Milliken: *The Interaction of the Visual and Thermal Environments on the Thermal Comfort and Acceptance of Indoor Space*. Report No. 80-04, Institute for Environmental Research, Kansas State University, Manhattan, KS (1980).
14. For a discussion of adaptation level theory, see Helson: *Adaptation Level Theory: An Experimental and Systematic Approach to Behavior*. Harper and Row, New York (1964).
15. American Society of Heating, Refrigerating, and Air-Conditioning Engineers: *ANSI/ASHRAE Standard 55-1981*. Atlanta (1981).
16. American Conference of Governmental Industrial Hygienists Committee on Industrial Ventilation: *Industrial Ventilation — A Manual of Recommended Practice*, 17th ed. Lansing, MI (1982).
17. Fanger, P.O.: *Thermal Comfort*, p. 97. McGraw-Hill, New York (1970).
18. Hall, J.F. and F.K. Klemm: Thermoregulatory Responses in Disparate Thermal Environments. *J. Appl. Physiol. 23*:540-544 (1967).
19. Boje, Nielson and Oleson: Article written in Danish, but is discussed by Fanger, P.O.: *Thermal Comfort*, p. 96 Op cit.
20. Carlton-Foss, J.A.: *Personality Factors in Thermal Acceptability and Comfort*. Ph.D. Dissertation, Saybrook Institute (1981).
21. Shapiro, D.: *Neurotic Styles*. Basic Books, New York (1965).
22. Wells, C.E.: The Hysterical Personality and the Feminine Character: A Study of Scarlett O'Hara. *Comp. Psych. 17*:353-359 (1976).
23. Carlton-Foss, J.A.: *Building Diagnostic Questionnaires*. Human-Technical Systems, Lincoln Center, MA (1980).
24. Selye, H.: *The Stress of Life*. McGraw-Hill, New York (1956).
25. Selkurt, E.E.: *Physiology*, 4th ed., Chap. 2. Little Brown, Boston (1976).
26. *The Merck Manual*, 13th ed., p. 1872. R. Berkow and J.H. Talbott, Eds. Merck Sharpe and Dohme Research Laboratories, Rahway, NJ (1977).
27. Schacter, S. and J. Singer: Cognitive, Social, and Physiological Determinants of Emotional State. *Psychol. Rev. 69*:379 (1962).
28. Frank, J.D.: *Persuasion and Healing*. Schocken, New York (1974).
29. Meninger, K.: *The Vital Balance: The Life Process in Mental Health and Illness*. Viking, New York (1963).
30. Vaillant, G.E.: Theoretical Hierarchy of Adaptive Ego Mechanisms. *Arch. Gen. Psych. 24*:107-118 (1971).
31. Vaillant, G.E.: *Adaptation to Life*. Little Brown, Boston (1977).
32. *Ibid*, p. 384.
33. Sullivan, R.: Patients' Deaths Tied to Heat are Investigated. *New York Times*, p. B1 (July 17, 1981).
34. Rotton, J.: *Correlations Between Climate and Human Behavior*. Paper presented at the Annual Meeting of the American Psychological Association, Washington, DC (1982).
35. Bridger, C.A. and L.A. Helfand: Mortality from Heat During July 1966 in Illinois. *Intl. J. Biometeorol. 12*:51 (1968).
36. National Research Council: *Airborne Particles*, p. 168. University Park Press, Baltimore (1979).
37. Baker, D.: Personal communication (1980).
38. Rohles, F.H. and W.V. Wells: Interior Design, Comfort, and Thermal Sensitivity. *J. Interior Design, Ed. and Res. 2, Part II*:36-44 (1979).
39. Rohles. F.H.: Temperature or Temperament: A Psychologist Looks at Thermal Comfort. *ASHRAE Part I*:541-551 (1980).
40. Baker, D.: Personal communication (1980).
41. Findlay, B.: Personal communication (September 1983).
42. Int-Hout, D.: Tight Building Syndrome — Is it Hot Air? *Heating, Piping and Air Conditioning* (January 1984).
43. Findlay, B.: Personal communication (September 1983).
44. Carlton-Foss, J.A.: *Psychological Factors in the Impacts of Indoor Air Pollution on Humans*. Paper presented at the Engineering Foundation Conference, *Management of Atmosphere in Tightly Enclosed Spaces*, Santa Barbara, CA (October 16-21, 1983).
45. Janssen, J.E.: Ventilation for Control of Formaldehyde in Mobile Homes. *ASHRAE Trans. 89*:511-515 (1983).
46. Hamilton, A. and H.L. Hardy: *Industrial Toxicology*, 3rd ed., pp. 321-326. Publishing Sciences Group, Acton, MA (1974).

47. Weiss, B.: Behavioral Toxicology and Environmental Health Science: Opportunity and Challenge for Psychology. *Am. Psychol. 38(11)*:1184 (November 1983).

48. Carlton-Foss, J.A.: Effects of "Ergonomic" Chairs on Human Performance at Word Processing. To be submitted to *Human Factors* (1984).

49. Spengler, J.D., R.D. Treitman and T. Tosteson: *Personal Exposure to Respirable Particles: Tale of Two Cities*. Paper presented at the *International Symposium on Indoor Air Pollution, Health, and Energy Conservation*, Amherst, MA (1981).

50. Ware, J.H., D.W. Dockery, A. Spiro et al: Passive Smoking, Gas Cooking, and Respiratory Health of Children Living in Six Cities. To be published in *American Review of Respiratory Diseases*.

51. Doll, R. and R. Peto: Mortality in Relation to Smoking: 20 Years' Observations on Male British Doctors. *Brit. Med. J. 2*:1525-1536 (1976).

52. Kellerman, G., C.R. Shaw and M. Luyten-Kellerman: Aryl Hydrocarbon Hydroxylase Inducibility and Bronchogenic Carcinoma. *New Eng. J. Med. 289*:934-937 (1973).

53. Brandenburg, J.H. and G. Kellerman: Aryl Hydrocarbon Hydroxylase Inducibility in Laryngeal Carcinoma. *Arch. Otolaryng. 104*:151-152 (1978).

SPECIAL VENTILATION PROBLEMS

Commercial/institutional indoor air quality study by the Bonneville Power Administration

ROBERT ROTHMAN
Environmental Specialist, Bonneville Power Administration, U.S. Department of Energy

The Bonneville Power Administration (BPA) is working to implement a region-wide conservation program. A part of that program proposes to retrofit commercial and institutional buildings in a manner that will reduce or minimize excess ventilation within a structure. However, if ventilation retrofits are to be pursued, BPA is concerned about the potential effects these measures could have on indoor air quality (IAQ). Since "acceptable IAQ" is an undefined term, and there is a paucity of available information documenting existing IAQ conditions, there are many questions that need to be answered. For this reason, BPA has undertaken a survey study to characterize IAQ conditions in relation to ventilation rates. With this information, BPA will be in a better position to assess the effects of reduced ventilation on IAQ.

The survey will consist of approximately 40 buildings located in the Portland, Oregon and Spokane, Washington vicinities. Most buildings will be offices, however, some libraries and schools will also be included.

Pollutants to be monitored include particulates, water vapor, formaldehyde, radon, carbon dioxide, and carbon monoxide. Carbon monoxide will be monitored only in those buildings with combustion sources.

Monitoring for radon, water vapor, and formaldehyde will be performed using passive monitors. These devices work on the principle of passive air diffusion and provide an integrated average concentration of the target pollutant. Particulate samples will be collected with a vacuum pump system on a filter, and will be gravimetrically analyzed to determine the concentration of respirable size particulates. Samples will be collected on a continuous basis and will provide an integrated sample. Several samples from each building will have approximately 16 polyaromatic hydrocarbons (PAHs) identified. Carbon dioxide and carbon monoxide will be sampled on a real-time basis, allowing for the identification of peak concentration periods. The obvious disadvantage of continuous passive sampling procedures is the absence of an episodic diary of pollutant occurance. The principle advantage is the relative small cost to collect an abundance of samples. Within each building, 7 to 16 sampling sites will be established.

In addition to monitoring pollutant levels, the study will include a ventilation test assessment. All buildings will have typical ventilation rates quantified using sulfur hexafluoride (SF_6) tracer gas. This type of ventilation measurement involves seeding the ventilation system with the tracer gas and subsequently monitoring the SF_6 gas decay rate using an electron capture gas chromatograph. Also, as many as 30 of the buildings will be monitored with a passive tracer gas technique using perfluorocarbon tracer gas (PFT). This method involves the passive release of PFT from multiple sources. The PFT assumes equilibrium within the structure and is collected on passive samplers by air diffusion through capillary adsorption tube. By analyzing the amount of PFT released compared to the amount adsorobed on the passive sampler and taking into account the volume of heated air space, an average air exchange rate in the structure can be calculated. This method is relatively new and not well tested in large commercial structures. The Bonneville Power Authority will help establish the feasibility for this methodology in large structures.

Cross contamination and entrainment

RICHARD W. GORMAN, CIH
National Institute for Occupational Safety and Health

Introduction

Indoor, non-industrial environments, such as office spaces, have been commonly thought of as being clean and trouble free environments. Although there always seems to have been, and still is, an ongoing battle to maintain temperature and relative humidity within comfortable ranges, build up of air contaminants has not been a particular concern. That is, not until rising energy costs has precipitated substantial changes in the manner in which buildings are designed or renovated and in the control strategies used to operate the heating, ventilation and air conditioning (HVAC) systems. The primary goal of these efforts is to reduce energy consumption and therefore minimize operating costs. Approximately 38% of the primary energy consumed by this nation is used to heat and cool buildings.[1] Forty percent of this, or 15% of the energy, is used to heat or cool office-type buildings. At a yearly cost of $2 to $4 to condition each cubic foot per minute (cfm) of air, it is understandable that energy conservation is given much attention.[2]

In addition to maintaining temperature and relative humidity within a comfortable range, we are all becoming aware of the importance of adequate outside ventilation. Complaints from a large percentage of building occupants of vague and non-specific illnesses where symptoms get worse as the work day progresses but subside after work and on weekends offer a clue that perhaps there are problems with the control of temperature, relative humidity and/or outside air ventilation. These three parameters are considered so important it is likely that a majority of the office-building problems would be solved if they are given due consideration. If complaints persist even after these important parameters are determined to meet currently accepted criteria, investigators then need to carefully evaluate the office area for the presence of airborne contaminants. These contaminants may be generated within the area, such as sometimes found in new buildings or recently renovated buildings, or outside of the building and brought into the area by either cross contamination or entrainment. Cross contamination and entrainment are two unique ventilation problems either, or both, of which may be contributing to the problem.

Cross contamination

Cross contamination occurs when a contaminant is present in an area of a building other than where it is generated and finds its way there without leaving the building.

Conditions conducive to cross contamination include:

1. Dual or multi-use buildings.
2. Office area and contaminant-generating area on safe HVAC system.
3. Separate HVAC systems serving adjacent areas, one of which is a contaminant-generating area.
4. Change in space usage relative to original design.
5. Heat wheel energy recovery device.

Cross contamination is most often associated with dual or multipurpose buildings where production or laboratory activities are under the same roof with offices or classrooms. A common HVAC system can re-circulate contaminants to the office areas causing complaints even though the resulting concentration is far below applicable occupational exposure criteria. Occasionally, an air cleaning device may be included in the return duct. If this is the case, there should be monitors to gauge the continuing effectiveness of the unit. If these monitors are missing or broken, higher than expected levels of the contaminant(s) may be crossing over to the office areas.

"Adjacent-area-pickup" can occur if there are malfunctions or poor design in HVAC systems that result in a negative pressure in a space adjacent to one which is a contaminant-generating area. While, except in special cases, the airflow should be from the clean area to the contaminated area, the malfunction or poor design may reverse this flow and allow contaminants to be picked up and circulated to non-contaminant-generating areas causing complaints.

A change in space usage may also precipitate complaints if an area that was originally designed to be used for offices or storage is converted to another function that generates a contaminant(s), i.e., a photography laboratory or art studio. Office areas on a common HVAC system may then be affected by strange chemical odors not previously noticed.

Contaminants may be transferred to areas they are not wanted via movement of air currents such as may be established by a "chimney-effect" where warm, contaminated air tends to rise. Elevator shafts, stairwells, dumb waiters and utility shafts can all be potentially important pathways in this regard.

Whenever there is a need for "once-through" or 100% outside air ventilation, some type of an energy reclaim system is advantageous. Units that incorporate liquid-to-air principles in a heat exchanger, like an ethylene gylcol system, offer little opportunity for cross over from the exhaust side to the outside air (supply) side of the HVAC system. However, air-to-air heat exchangers, such as rotary wheel exchangers (Figure 1), may allow cross over even when operating correctly. The rotary, air-to-air heat exchanger, popularly called a heat wheel, is a revolving cylinder made up of axially air-permiable media having a large internal surface for intimate contact with the air passing through it. The air duct connections are arranged so that each of the air streams (exhaust and outside air) flow axially through approximately one half of the wheel in a counter flow pattern. In Figure 1, warm return air from the building is collected and fed to the heat recovery system in compartment A, flows down through the heat recovery wheel into compartment B and exits the building. Supply fans bring 100% outside air into the heat recovery system in compartment C where it is filtered, preheated, sent through the other half of the heat wheel and delivered to the building. The porous media that is heated from the warm duct (exhaust) air stream, rotates into the colder duct (outside air) air stream where it releases the newly obtained energy. A purge system can be added to minimize cross over of entrained air but, for the purge to function properly, the exhaust side of the wheel must always be a lower pressure than the fresh air side. Any violation of this condition would result in an additional cross over. This type of an energy reclaim device would not be appropriate for situations where highly toxic or extremely odorous agents are handled.

Entrainment

Entrainment occurs when contaminants are brought into the building from the outside. Conditions conducive to entrainment include:

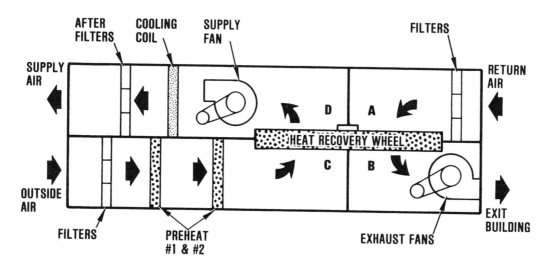

Figure 1 — Heat recovery wheel.

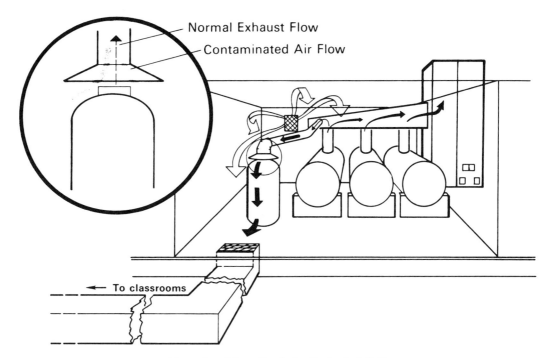

Figure 2 — Schematic of school's physical plant.

1. Fresh air intakes that are located close to or down wind of exhaust stacks.
2. Exhaust stacks too low to discharge outside the turbulent wake caused by airflow over and around the building.
3. Buildings under negative pressure.

A classic example of entrainment would be the situation where contaminated air exits from an exhaust stack only to be captured and returned to the building via a nearby fresh air intake. On these circumstances, we find that an architecturally pleasing design may be at odds with other designs that minimize the potential for reentry. For example, while a 25 foot exhaust stack may be preferable, they are usually aesthetically unacceptable.

Building or parts of buildings under negative pressure with respect to the outside can cause numerous potential problems. In addition to creating a situation whereby outdoor contaminants can be drawn into the building, this condition can result in cold drafts, decreased hood efficiencies, increased housekeeping, wasted energy due to fan inefficiency, and back-draft or "draw-down" on combustion stacks.

The potential problems caused by negative building pressure is illustrated by a recent health evaluation[3] at a school where students and staff were complaining of headaches, mucous membrane irritation and skin rashes starting with the first heating season after completion of an energy conservation program. The program included the application of polyurethane foam in the outside walls and on the roof, inside storm windows, the sealing of outside air vents, and installation of a new water heater. Forty exhaust fans were being operated with inadequate makeup air; therefore, the school building was under negative pressure. There was no allowance for outside air ventilation in the classrooms. In addition to the lack of adequate outside air ventilation, boiler gases were entering the boiler room via the flue on the new water heater (Figure 2) and distributed throughout the school via a utility tunnel (Figure 3) due to the draw placed on the tunnel by the negative building pressure. There has been no recurrence of the problem since outside air ventilation has been introduced and reentry of boiler gases via the water heater flue corrected.

It is possible for the lower floors of a building to be under negative pressure and the top floors to

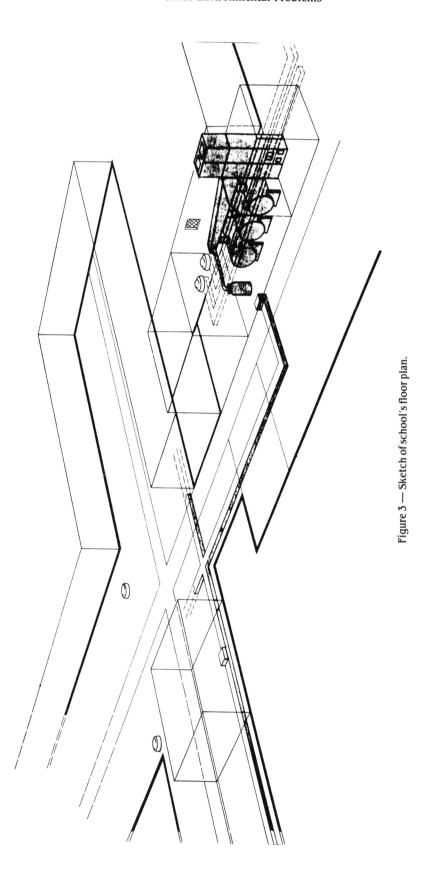

Figure 3 — Sketch of school's floor plan.

be under positive pressure due to the "chimney-effect" where warm air tends to rise pressurizing the upper floors. If the pressure in the space served exceeds the optimal pressure used in the design of the supply fan side of the HVAC system, then the supply fan may be overworked and inefficient, and the amount of air delivered to the office space may be less than expected.

Buildings nearby or downwind of other buildings that are exhausting offensive contaminants, whether from a chemical or a combustion process, are particularly susceptible to the problems caused by entrainment. Routes of entry can be the outside air intake of the HVAC systems or openings in the building skin, i.e., windows, doors or cracks left unsealed during construction. Entrainment is not only of concern in an industrialized setting. The same routes of entry previously mentioned can also bring allergens such as pollen, dust and molds into a building. Therefore, rural environments do not preclude the need to consider entrainment of contaminated ambient air. Obviously, entrainment of exhausted contaminants present the same problems in rural settings.

Evaluation considerations

Five important issues to consider when evaluating office environmental problems include:

1. Temperature.
2. Relative humidity.
3. Outside air ventilation.
4. Contaminants generated within the area.
5. Contaminants generated outside the area and brought into it (cross contamination an entrainment).

Any one, or combination, of these may be responsible for the problem. So, the question is where do you start? At this point, input from the work force can be very useful in determining the character and scope of the problem. A brief questionnaire administered as soon as the problem develops can provide useful information on symptoms and help to determine how much of the building is being affected. The longer the problem persists the more confused workers become about whether the symptoms are theirs or their co-workers who have been frequently complaining for some extended period of time. The initial questionnaire should be short and simple to complete and score. The same questionnaire can be used to determine the success of any corrective actions.

Evaluation methods

Issues 1 through 4 have been discussed during earlier presentations. Methods useful in evaluating the fifth issue, cross contamination and/or entrainment include:

1. Subjective (odor recognition).
2. Standard industrial hygiene techniques.
3. Tracer gas such as sulfur hexafluoride (SF_6).

A contaminant that has a characteristic odor, i.e., diesel fume or solvents can be more easily traced to the origin. It is also possible to use an odorous agent such as banana or winter-green oil as a tracer gas by releasing it in one area and trying to detect it in the office area in question.

Standard industrial hygiene sampling techniques can be useful, especially if a particular contaminant is suspected of causing the problem. Air sampling strategies have been designed to test for the presence of a class of agents such as organic vapors and aldehydes, but the results are often difficult to interpret as they are usually well below those levels expected to be of any health significance. Interpretation of air sampling data may be enhanced by comparing the results from an area where there are complaints to the results obtained in an area in the same building where there are no complaints. A low concentration of an airborne contaminant in an office area may elicit complaints whereas, the same concentration in the area where it is generated may be tolerated simply because it is associated with the activities there.

Cross over or entrainment via a fresh air intake can be quantitated using a tracer gas such as sulfur hexafluoride (SF_6). However, this technique is expensive because of the high cost of the pure gas and the analytical equipment needed to measure it. Cross over situations where dilution factors are 100 or 1000 can be studied by releasing a known concentration of SF_6 and measuring for it in the problem office area using a portable IR detector. To study entrainment or reentry via fresh air intakes where dilution factors may be 10^4 to

10^5, a more sensitive detector, such as an electron capture gas chromatograph is needed.[4]

Corrective actions

Once the contaminant(s) potentially responsible for the problem is identified and the generation location pin-pointed, corrective action amounts to interrupting the pathway, whether it be a cross contamination or entrainment mechanism. Possible corrective actions include:

1. Rebalancing HVAC systems.
2. Local exhausting of contaminants at the source.
3. Isolation of contaminant-generating areas.
4. Raise exhaust stacks.
5. Insulate and seal cracks.
6. Provide adequate makeup air.
7. Maintain a well-trained maintenance staff.
8. Substitution of offending agent.

Conclusions

The control strategies and equipment used in HVAC systems are becoming more and more sophisticated. Those individuals who understand the system best (design and construction engineers) often leave after a project is completed. Individuals left to maintain the systems, even though trained to some extent, are not always knowledgeable enough to trouble-shoot the systems as conditions change.

A recent article in the *Cincinnati Enquirer* reported on a survey of 448 new building owners. More than half reported problems with their HVAC systems and more than a third of the owners have resorted to litigation or formal arbitration.

It is evident that new HVAC systems do not continue to operate as designed and some are poorly designed. Investigators, such as industrial hygienists, for the most part, do not have the training to properly evaluate these new systems. This symposium has reenforced this issue. I hope that it also provides the impetus to initiate specific training courses on the trouble-shooting of building ventilation systems so that those charged with the evaluation of office environmental problems will be better equipped for this challenge.

References

1. *A New Prosperity.* Brick House Publishing Company, Hanover, MA (1981).
2. McNall, Preston, Ph.D.: Personal communications. Senior Research Engineer, National Bureau of Standards, Building 226, Washington, DC 20234; (301) 921-3447.
3. National Institute for Occupational Safety and Health: *Wappingers Falls Central School District.* Hazard Evaluation and Technical Assistance Report. HETA 83-172. Cincinnati, OH (1984).
4. Hampl, V. and R. Hughes: *Tracer Gas Technique — A Useful Method for Indoor Air Pollution Evaluation.* Presented at the American Industrial Hygiene Conference, Philadelphia, PA (1983).

Case presentations: problems caused by moisture in occupied spaces of office buildings

PHILIP R. MOREY
National Institute for Occupational Safety and Health

Industrial hygiene studies in two office buildings are described. Both buildings have had environmental problems involving moisture incursion into occupied space, and both are characterized by heating, ventilation and air conditioning (HVAC) systems which supply inadequate amounts of outdoor air to occupants. Levels of airborne microorganisms in some offices in these buildings were as much as 10 to 100× higher than that present in the outside environment. In the opinion of the author, the guideline of an airborne viable particulate level in excess of 1000 colony forming units (CFU)/m^3 as an indicator of an office environment in need of improvement may be useful in future aerobiological studies.

Introduction

The Division of Respiratory Disease Studies of the National Institute for Occupational Safety and Health (NIOSH) through its Health Hazard Evaluation Program is often requested by other governmental agencies to evaluate illnesses associated with the office workplace. A common problem found in many office workplaces is tight building syndrome.[1-3] Complaints encountered in tight buildings include mucous membrane irritation, fatigue, headache, and concerns about inadequate or stuffy conditioned air. Tight building syndrome may be caused by inadequate make-up air, and air contaminants such as vapors from building adhesives and office equipment.[1-4] In several office workplaces recently studied by NIOSH, symptoms compatible with hypersensitivity pneumonitis (HP) or humidifier fever have also been reported.[5] These illnesses are caused by a broad array of microbial agents.[3,5,6] It has been suggested that an airborne viable particulate level greater than 1000 CFU/m^3 may be indicative of an office environment in need of improvement.[5]

Described in this paper are environmental studies of two office buildings where some complaints given by occupants are compatible with tight building syndrome and other symptoms are suggestive of HP-like illnesses. One of these buildings has been described elsewhere in these proceedings (Building E in reference 5), and has had a long history of problems associated with the proliferation of mold in the indoor environment. The second building (Building F) experienced a massive flood that raised concerns about the possible presence of biological agents in occupied spaces. These case history presentations offer the industrial hygienist background information to be used for determining if an office environment is contaminanted with microbial agents.

Building E

This office complex located in an eastern city houses over 2000 employees, was constructed in 1969, and consists of three interconnecting structures. A 15-story office tower is joined to a 7-story building by a 4-story structure. Thus, the lower four floors of the building complex are interconnected. Persistent environmental problems have been experienced within the building including floods from roof leaks and a relative humidity that often exceeds 70% during the summer air conditioning season. HP was diagnosed by the private physician of one building occupant. Complaints of shortness of breath, persistent cough, unexplained fatigue, and eye, nose and throat irritation have been reported by occupants, especially those located on lower floors of this office complex.

Description of HVAC system and air flow measurements

The HVAC system of Building E is unique in that air is not recirculated; the central HVAC system supplies 100% outdoor air to occupied space. This is accomplished by seven main air handling units (AHUs). AHUs are located in upper floor and roof top areas and are supposed to operate from 0600 to 1730 hours on Mondays through Fridays. Each main AHU contains a filter bank (no rated ASHRAE atmospheric dust spot efficiency),[7] heating and cooling coils, and a fan. Conditioned air leaving each AHU is conducted through one or more shafts to common supply plenums formed by the suspended ceiling and the slab of the floor above. Supply air moves through the pressurized plenum (no ducts) and enters occupied space through diffusers located in the suspended ceiling. Some air is removed from Building E by restroom exhausts. Additional air enters grilles or registers in main corridors, moves through a series of shafts and is exhausted into an underlying parking garage.

Air within Building E is further conditioned and recirculated by 393 small AHUs and 1096 peripheral fan coil units (FCUs). Small AHUs are found in interior rooms and are operated by occupants. Air from occupied space enters each small AHU through a grille located in its lower front. Air is filtered (no rated efficiency), cooled (cooling coils only), and recirculated to a cluster of offices through a duct with a number of vertically oriented supply registers. Peripheral FCUs are located along outside walls, are controlled by room occupants, and condition air within occupied space by heating or cooling. Each unit contains a filter (no rated efficiency), heating and cooling coils, and a fan. There is no provision for incorporation of outdoor air into peripheral FCUs.

During a visit to Building E in January 1983, it was observed that fans of main AHUs were turned off for several hours or more during the working day. This practice occurs during periods of cold winter and warm summer weather because of the inability of heating and cooling coils to adequately temper outdoor air.

That outdoor air provided by main AHUs was adequately distributed to occupied space on the lower floors of Building E was questionable. The lower surface of the suspended ceiling-supply plenum in many office zones was broken by absent, damaged, and improperly fitted tiles. Sometimes ceiling tiles were intentially propped up or improperly fitted (by room occupants) so as to direct air flow into a particular office. Ceiling diffusers were absent in some offices and thus these rooms had no direct source of outdoor air.

Air flow measurements were made in one office suite which was composed of 12 floor-to-ceiling rooms, two wide corridors, and one narrow hallway (Table I). Fourteen people worked in this zone. At a time when main AHUs were operating, the amount of air entering this office suite through individual diffusers varied from 1.87 to 12.5 liters per second (L/s) (3.75 to 25 cubic feet per minute (cfm)) for a total of only 43 L/s (86 cfm). This was equivalent to about 3 L/s (6 cfm) of air per occupant or only about 30% of the recommended minimum amount[8] for an office building where cigarette smoking is permitted. Some of the diffusers in this office suite were attached to short length (0.5 m) of metal ductwork that ended blindly within the suspended ceiling plenum. Airflow through these diffusers was lower than that through diffusers without attached ductwork (Table I).

These observations collectively indicate that Building E is inadequately ventilated. It is doubtful that outdoor air supplied by main AHUs is adequately distributed to building occupants. Since conditioned outside air is not transported to diffusers by ducts, breaks in the integrity of the lower surface of the supply air plenum, as was commonly observed, will cause a redirection of the flow of outdoor air preferentially into some zones at the expense of all others. In the one office suite where ventilation measurements were made, main

TABLE I
Air Supplied to Office Suite in Building E Through Ceiling Diffusers

Diffuser No.	Air Flow Through Diffuser[a]	
	L/s	cfm
1	12.5	25.0
2[b]	3.75	7.5
3[b]	3.75	7.5
4[b]	3.75	7.5
5[b]	1.87	3.75
6[b]	3.12	6.25
7	5.0	10.0
8	7.5	15.0
9[b]	1.87	3.75

[a]The amount of conditioned air entering occupied space through ceiling diffusers was the product of the average air velocity (Datametrics* hot wire air velocity meter) and the area of the diffuser inlet in the suspended ceiling. Because of their irregular shape, the louvered grilles were removed and air velocity measurements were made directly at the inlet in the suspended ceiling. Thus, these air flow measurements may be somewhat greater than actual values. (* Mention of company names or products does not constitute endorsement by the National Institute for Occupational Safety and Health.)

[b]Duct, 0.5 m long attached to these diffusers. Ducts end blindly in ceiling plenum. The inlet area for all diffusers except No. 1 is 225 cm^2 (0.25 ft^2). The inlet area for diffuser No. 1 is 450 cm^2 (0.5 ft^2).

AHUs supply only about 3 L/s (6 cfm) of outdoor air per occupant. Of course, when main AHUs are not operated the only outdoor air reaching occupants will be that which infiltrates the building envelope.

Microbial studies

Sources of microbial contamination in Building E are described in detail elsewhere[5] and are briefly summarized as follows:

1. The relative humidity in occupied space may exceed 70% during the summer air conditioning season.

2. Some masonry ceiling slabs are permeated with water. Some suspended ceiling tiles are so wet that slight upward finger pressure is sufficient to push right through tile material. Wet tiles may be covered with a thick fungal mycelium. Condensate water overflowed drain pans in small AHUs and FCUs causing floods in some offices.

3. Drain pans of main AHUs and FCUs contained stagnant water and slime. The bacterial content in stagnant water in drain pans of main AHUs and FCUs was approximately 1×10^7 per ml. Drain pans of small AHUs are totally enclosed in interior walls making preventive maintenance impossible.

4. AHUs and FCUs are sources of microorganisms. Filters act as microbial reservoirs since they are not replaced with adequate frequency. The dust present in a filter from one small AHU contained about 3×10^7 viable fungi per g. Sound lining on the inside surfaces of some small AHUs and FCUs was encrusted with mold.

Air sampling for fungi was carried out in offices in Building E in January 1983 at a time when outside concentrations varied from 50 to 500 CFU/m^3. The air in most offices contained numbers of fungi less than 200 CFU/m^3. However, sampling in several interior rooms where conditioned air was being provided by small AHUs showed that levels of fungi varied from about 200 to 1500 CFU/m^3.

A second survey was conducted in Building E in September 1983. At that time levels of fungi in outside air were approximately 800 CFU/m^3. Fungal sampling was carried out in the stagnant air in a room that had at one time been occupied by a complainant. Airborne counts exceeded 7000 CFU/m^3 (Table II). The drain pan of the FCU serving this office contained stagnant water and slime. When the unit was turned on, and was agitated (as if backing an office chair into its metal housing) levels of fungi rose to > 94,000 CFU/m^3. Air sampling was subsequently carried out in a second room now occupied (without complaint) by the individual who had previously resided in the first office (Table II). The drain pans of the two FCUs in the second office were dry. Levels of airborne fungi recovered on two occasions in this office were low (less than outdoor concentrations). On one occasion when the occupant used a portable electrostatic precipitator, the concentration of airborne fungi recovered was < 25 CFU/m^3.

Additional air sampling was carried out in an interior room where conditioned air was being supplied through two ceiling diffusers. Several suspended ceiling tiles in this office were wet and covered with mold. The concentration of airborne fungi in this office was about 9000 CFU/m^3 (Table II).

FCU in many offices in Building E contained sound lining encrusted with mold and debris. Sampling carried out in one of these office (Room 4, Table II) showed that large numbers of fungi can be released from FCUs. Dense spore clouds were also disseminated into occupied space by the small AHUs that provide conditioned air to interior rooms in Building E (see Table IV, Reference 5).

Recommendations

Remedial actions suggested for Building E include the following:

1. Promptly and permanently prevent moisture incursion into occupied space such as from roof leaks and drain pan overflows.

2. Initiate a preventive maintenance program to remove slime from main AHUs and FCUs. Install access panels in small AHUs so that the drain pan can be subjected to a preventive maintenance program.

3. Upgrade the preventive maintenance program for FCUs and small AHUs. Replace dirty interior sound lining in these units. Change filters more frequently. A vacuum

TABLE II
Airborne Fungi in Building E Offices Where Conditioned Air is Supplied by Perimeter Fan Coil Units or Through Diffusers in the Suspended Ceiling

Description of Room and HVAC System Component	CFU/m³*
Room No. 1 Had been occupied by a complainant; FCU off; stagnant water in drain pan; foul odor; unoccupied; second floor exterior room	7,360
Room No. 1 As above except that FCU on and unit agitated; result obtained on two successive days	>94,000
Room No. Occupied without complaint by individual formerly in Room No. 1; FCUs off; electrostatic precipitator off; second floor exterior room.	2 165
Room No. 2 As above; one FCU on; electrostatic precipitator operating**	<25
Room No. 3 Second floor interior room with one occupant; suspended ceiling contains two diffusers; mold growing on several ceiling tiles	9,025
Room No. 4 Exterior room on 13th floor; FCU on; unit agitated; no water in drain pan; one occupant	14,290

* Level of fungi in outdoor air = 800 CFU/m³. Aerobiological sampling techniques described in reference 5.
** Unit consists of prefilter, electronic precipitator, charcoal filter, and a fan.

incorporating a high efficiency particulate air (HEPA) filter should be used to clean accumulated debris from FCUs and small AHUs.

4. Keep the relative humidity in occupied space below 70% during the summer air conditioning season.

5. Replace damaged and missing suspended ceiling tiles. The integrity of the lower surface of the positive air supply plenum must be maintained.

6. Main AHUs should at a minimum be operated at all times when the building is occupied. Outdoor air should be provided at a minimum of 20 cfm per occupant.

7. Conditioned air would be better distributed to occupants if individual offices were of the open concept type and not the floor-to-ceiling type as commonly found on lower floors of Building E.

Building F

This 18-story office building, located in an eastern city, was constructed in 1972, and houses about 1900 occupants. In November 1982, plumbing lines serving a heat pump located above the suspended ceiling of a second floor office ruptured and caused a massive flood on the second, first and mezzanine floors. Carpet in flooded zones was subsequently vacuumed and treated with a germicidal chemical.

In 1983, NIOSH was asked to evaluate environmental conditions in Building F. A brief walk-through survey was carried out in June of that year. This was followed in December 1983 by a survey involving extensive microbial sampling. Complaints offered by occupants throughout the building included headaches, eye and throat irritation, and a perception that indoor air was stuffy or poorly ventilated. Employees located in zones that had been flooded in November 1982 reported chills and flu-like symptoms. This could not be documented by absentee records because ill persons continued to work. After the major flood of November 1982, occupants throughout the building developed an increased perception that the quality of their working environment was poor.

Building ventilation

Air within Building F is conditioned by over 360 heat pumps (about 20 per floor) located in the plenum above the suspended ceiling. Conditioned air from heat pumps is ducted to zones and enters occupied spaced through diffusers located in the suspended ceiling. Air from occupied space returns to the ceiling plenum through slots around light fixtures, and then is reconditioned by heat pumps. Some air from the ceiling plenum on each floor is removed from the building through a single riser that terminates in plenums housing rooftop exhaust fans. Additional air is removed from the building by restroom and cafeteria exhausts (fans on roof). Make-up air is ducted into several locations within the ceiling plenum on each floor from risers that connect with rooftop plenums housing supply fans.

Heat exchange between the outside environment and building heat pumps is accomplished by evaporative condensers located outdoors on the 17th floor. Heat pumps operate in a closed loop system with pipes (water containing a corrosion inhibitor in pipes) in each unit continuous with pipes that pass through evaporative condensers. Water in evaporative condenser reservoirs never mixes with water in the closed loop-heat pump system. The flood on the lower floors of Building F in November 1982, was caused by a rupture in the closed loop system serving heat pumps.

Two sanitary vents, and one restroom and one cafeteria exhaust stack were located on the roof within 3 meters of intake louvers where outdoor air enters the plenums housing air supply fans. One of the sanitary vents was less than one meter from a louvered outdoor air intake. Contamination of make-up air with exhausts from vents and stacks occurred whenever fans supplying make-up air to the building were operated.

It was observed that fans providing make-up air in Building F are shut down whenever the outside temperature is below -1°C (30°F) or above 32°C (90°F). This action was taken because of the inability of the HVAC system to temper very cold or very warm outdoor air. During December when one of the two supply fans providing make-up air to the building was shut down for repair, levels of carbon dioxide (CO_2), as measured by short-term indicator tubes, were equal to or greater than 800 ppm in many offices including some on the second floor. A CO_2 level of 1500 ppm was recorded in one second floor office. It has been suggested that elevated levels of CO_2 (600 to 1000 ppm, and higher) are indicative of an inadequate supply of outdoor air in occupied space and are associated with discomfort.[4,9] In Building F, where mechanical systems providing make-up are often shut down, complaints can be expected from occupants coinciding with elevated levels of CO_2 in the air being recirculated in occupied space.

Microbial studies

Sampling for airborne mesophilic and thermophilic fungi and for airborne bacteria was carried out in Building F in December 1983 (Tables III and IV). The average number of mesophilic fungi in air samples from locations that had been flooded (second floor, mezzanine) was 10,200 CFU/m^3. Levels of fungi in air from non-flooded floors were about 3000 CFU/m^3, whereas average outdoor concentrations were 1290 CFU/m^3. The variable number of fungi recovered within any type of location (std. dev. is equal to or $>$ than mean, Table III) was likely the result of collecting samples over a three-day period from different building locations. In spite of the variability noted, statistical analysis showed that levels of airborne mesophilic fungi in flooded locations were significantly greater than those found either in non-flooded locations or in the outdoor air.

Levels of airborne thermophillic fungi and bacteria found in flooded locations were also higher than those present in non-flooded zones (Tables III and IV). However, significance of difference between group means could not be demonstrated.

Results of sampling for airborne mesophilic fungi on a floor-by-floor basis in Building F are presented in Table V. Numbers of samples taken per floor are too meager to facilitate a statistical analysis. However, it was apparent that the highest

TABLE III
Concentration of Airborne Fungi in Building F Offices that had been Flooded, in Non-flooded Offices, and in Outdoor Air[a]

Location	Mesophilic Fungi, CFU/m^3 (Mean ± Std. Dev.)[b]	Thermophilic Fungi, CFU/m^3 (Mean ± Std. Dev.)[c]
Flooded	10,200 ± 11,400	3,090 ± 8,670
Non-flooded	3,020 ± 3,120	650 ± 510
Outdoor Air	1,290 ± 1,990	220 ± 380

[a]Mesophilic fungi = incubation at 28 to 30°C; thermophilic fungi = incubation at 42 to 45°C; sampling carried out from December 13 through December 15, 1983 by methods given in reference 5. Group means are an average of 10 to 15 samples.

[b]Mesophilic fungi: means of three groups are significantly different at the 1% level by one way analysis of variance; flooded versus non-flooded and flooded versus outdoor air group means are significantly different at the 5% level.

[c]Thermophilic fungi: no significant differences between group means.

TABLE IV
Concentration of Airborne Bacteria in
Building F Offices that had been Flooded,
in Non-Flooded Offices, and in Outdoor Air*

Location	Bacteria, CFU/m^3 (Mean, ± Std. Dev.)**
Flooded	4,440 ± 7,290
Non-flooded	1,920 ± 1,470
Outdoor Air	267 ± 254

* Bacteria collected from December 13 through December 15, 1983 by methods given in reference 5. Collection media = tryptic soy agar (100 µg cycloheximide/ml); incubation temperature = 37°C. Group means are an average of 10 to 15 samples.

** No significant differences between group means.

concentration of airborne fungi occurred on the second floor. Among the 11 air samples collected on that floor, three were from a small office, two were from a large office, four were from an agency library, and two were collected in the lobby near a bank of elevators (Table VI). Although the carpet found in all second floor and mezzanine locations had been flooded in November 1982 and all carpet was heavily worn and frayed, fungal counts where the highest in the two offices and library location (Tables V and VI). The reason for the high levels of fungi found in the three second floor locations is unknown.

Microbial sampling conducted in Building F showed that numbers of airborne mesophilic fungi were higher in flooded as compared to non-flooded offices. The concentration of fungi present in indoor air both in flooded or in non-flooded locations was higher than that found outdoors (Table III). Mechanical ventilation is usually associated with a lowering of the indoor concentration of airborne microorganisms as compared to levels encountered outdoors or in naturally ventilated (open windows) structures.[10-12] Since the levels of microorganisms recovered in Building F were higher than those outdoors, it may be concluded that an internal source(s) of fungi exists especially in offices that were flooded. That carpet might be a source of intramural fungi was suggested by preliminary data (unpublished) which shows that carpet from zones that had been flooded contains significantly more entrained fungi than carpet from non-flooded locations.

Recommendations

Remedial actions suggested for Building F include the following:

1. Replace carpet that has been flooded or initiate a preventive maintenance program involving regular vacuuming of carpet with an instrument incorporating a HEPA filter.
2. Fans supplying outdoor air should be operated at all times when the building is occupied. Outdoor air should be provided at a minimum of 20 cfm per occupant.
3. Exhausts from sanitary vents and other stacks must not contaminate outdoor air intakes. These vents and stacks should be raised or relocated.

Discussion

It has been suggested elsewhere that a level of viable particles in excess of 1×10^3 per m^3 indicates that the office environment may be in need of investigation or improvement.[5] Numbers of fungi up to 100× this level were found in some offices in Building E. The sources of microbial contamination in this building include slime in AHU and FCU drain pans, contaminated sound lining in AHUs and FCUs, continued moisture incursion from roof leaks and drain pan overflows, and an almost non-existant preventive maintenance program. Remedial action needed in Building E has already been discussed.

In Building F it is clear that in some formerly flooded zones, fungal counts exceed the suggested guideline of 1×10^3 viable particles/m^3 by a factor of 10 to 20×. Carpet may be a source of microbial contamination in the flooded locations of this building. Appropriate remedial action to improve

TABLE V
Comparison of Airborne Mesophilic
Fungal Levels by Floor in Flooded
and Non-Flooded Locations in Building F

Floor	Flooded (F) or Non-Flooded (NF)	CFU/m^3	Number Samples
Mezzanine	F	2,150	4
2	F	13,140	11
7	NF	3,510	2
11	NF	5,930	2
12	NF	3,550	4
13	NF	620	1
14	NF	950	2
16	NF	1,800	2

TABLE VI
Mean Concentration of Airborne Mesophilic Fungi Present in Locations on the Second Floor of Building F. All Locations had been Flooded One Year Earlier.

Description of Location	CFU/m^3	Number Samples
Small Office Room	12,320	3
Large Office Room	10,040	2
Library	21,300	4
Lobby Near Elevators	1,150	2

the quality of the environment in Building F should involve either replacement of carpet or removal of fungi from carpet. If as a result of flooding, fungi proliferated in the carpet, vacuuming with an instrument incorporating a HEPA filter should remove spores and other microbial products that may become airborne. A common feature of office buildings with HP-like illnesses may be moisture incursion into occupied space or into the HVAC system.[5] For this reason, regardless of whether carpet in Building F is replaced or cleaned, it is essential that additional flooding in occupied space be prevented.

References

1. Hicks, J.B: Tight Building Syndrome: When Work Makes You Sick. *Occup. Health and Safety*, pp. 51-56 (January 1984).
2. Salisbury, S., P. Roper, B. Miller and A. Kelter: 101 Marietta Tower Building. *Health Hazard Evaluation Report*, TA 80-122-1117. NIOSH, CDC, PHS, DHHS (1982).
3. Kreiss, K. and M.J. Hodgson: Building-Associated Epidemics. *Indoor Air Quality*, pp. 87-106. P.J. Walsh, C.S. Dudney and E.D. Copenhaver, Eds. CRC Press, Boca Raton, FL (1984).
4. Bell, S.J. and B. Khati: Indoor Air Quality in Office Buildings. *Occup. Health in Ontario* 4:103-118 (1983).
5. Morey, P.R., M.J. Hodgson, W.G. Sorenson et al: Environmental Studies in Moldy Office Buildings: Biological Agents, Sources and Preventive Measures. (*These Proceedings*).
6. Medical Research Council Symposium: Humidifier Fever. *Thorax* 32:653-663 (1977).
7. American Society of Heating, Refrigerating, and Air-Conditioning Engineers: *Standard 52-56: Method of Testing Air-Cleaning Devices Used in General Ventilation for Removing Particulate Matters*. Atlanta, GA (1976).
8. American Society of Heating, Refrigerating, and Air-Conditioning Engineers: *Standard 62-1981: Ventilation for Acceptable Indoor Air Quality*. Atlanta, GA (1981).
9. Rajhans, G.S.: Indoor Air Quality and CO$_2$ Levels. *Occup. Health in Ontario* 4:160-167 (1983).
10. Hirsch, D.J., S.R. Hirsch and J.H. Kalbfleisch: Effect of Central Air Conditioning and Meterologic Factors on Indoor Spore Counts. *J. Allergy Clin. Immunol.* 62:22-26 (1978).
11. Rose, H.D. and S.R. Hirsch: Filtering Hospital Air Decreases *Aspergillus* Spore Counts. *Am. Rev. Resp. Dis.* 119:511-513 (1979).
12. Solomon, W.R., H.A. Burge and J.R. Boise: Exclusion of Particulate Allergens by Window Air Conditioners. *J. Allergy Clin. Immunol.* 65:305-308 (1980).

A typically frustrating building investigation*

STANLEY A. SALISBURY
Regional Industrial Hygienist, NIOSH Region IV, Atlanta, Georgia

Introduction

I like to refer to this case study as a typically frustrating building investigation. Those of you who have conducted building studies will understand and appreciate my use of the term frustrating. After presenting my case study of the NIOSH 101 Marietta Tower investigation, perhaps the rest of you will appreciate the frustration industrial hygienists experience when conducting indoor air quality investigations.

On March 4, 1980, employees of the U.S. Public Health Service (PHS), Region IV, requested technical assistance from the National Institute for Occupational Safety and Health (NIOSH) in investigating the cause of health complaints reported by many staff members working on the 10th, 11th, and 12th floors of the 101 Marietta Tower Building, a 36 story high-rise located in downtown Atlanta. The employees were experiencing eye irritation (especially contact lens wearers), nasal congestion, and other sinus problems which they suspected were caused by their office environment.

During the spring and summer of 1980, NIOSH industrial hygienists, and medical officers from the Center for Environmental Health (CEH), Centers for Disease Control, interviewed the affected PHS employees. It was initially suspected that symptoms were the result of poor indoor air quality. As word of the investigation spread, similar complaints were received from other occupants of the building outside the PHS offices. It soon became apparent the problem was not confined to the 10th-12th floors. Because similar health problems were being reported with increasing frequency among other office workers throughout the country, in April, 1981, NIOSH and CEH decided to initiate an extensive environmental and epidemiologic investigation of the Marietta Tower building.

Several questions arose which we felt should be addressed:

1. Were the symptoms experienced on certain floors, or noted throughout the building?
2. What was the full range and severity of these symptoms?
3. Since all the initial complainants were women, were the symptoms predominantly among females?
4. Were clerical workers more susceptible than professionals?
5. Were the symptoms the result of exposure to some unknown air contaminant such as tobacco smoke, photocopying fluid vapors, or dusts from structural materials such as fireproofing insulation?
6. Were symptoms related to the amount of outdoor air ventilation supplied to the building?
7. Was the occurrence of symptoms effected by the source or quality of the outdoor air supplied to the building?

Background

The 101 Marietta building is typical of many modern high rise structures. The exterior is composed entirely of large glass panels which can not be opened. Although this a privately owned building, at the time of the NIOSH investigation, 29 of 35 floors were leased by the Federal Government to provide office space for 15 separate federal agencies employing a total of approximately 2500 people. The basement and ground floor were occupied by the U.S. Postal Service. Private tenants and Postal Service Employees were not included in the survey.

Floor plans and interior designs varied from floor to floor. Some areas were fully enclosed private offices and others were open office spaces with partial room dividers. Most floors were a mixture of both.

---

* Case Study — NIOSH Health Hazard Evaluation TA 80-122-1177, 101 Marietta Tower, Atlanta, Georgia.

The building's heating, ventilation, and air conditioning (HVAC) system is a typical chilled water system with 2 separate 5-zone, constant volume, variable temperature air handling units located on each floor. One air handler circulates air on the east half of the floor and the other circulates air to the west half of the floor. Chiller units are housed in the basement and the cooling tower is located on the top level of an adjacent parking garage. Air is supplied to office spaces from ceiling diffusers and perimeter slot vents above the window panels. Also typical of many newer HVAC systems, the space above the suspended ceiling is used as the return air plenum. Air is pulled into this space through slots around the flush mounted light fixtures. Return air is pulled into the air handler mechanical room where it is filtered, cooled, and recirculated back to office spaces through air supply ducts. Supply air temperature is regulated using thermostatically controlled zone bypass dampers. Outdoor air, supplied through a separate duct system, mixes with return air inside the mechanical room. A manually adjusted damper is used to set the outdoor air volume according to HVAC design specifications. Outdoor air for the upper 15 floors of the building is brought in through vents on the roof. The lower 20 floors receive outdoor air from intake vents located at the east and west end of a breezeway formed by the building and adjacent parking garage. The HVAC system is shut down overnight and during weekends and holidays.

Evaluation design and methods

The investigation had two components: 1) an epidemiologic investigation in representative areas of the building to identify affected employees, describe adverse health effects experienced, and identify risk factors for building occupants; and 2) an industrial hygiene survey in those areas to evaluate the effectiveness of HVAC systems, to measure the level of suspected air contaminants, and to generally assess environmental conditions within the building.

In April 1981 we initiated a questionnaire survey of the federal employees working in the building while concurrently conducting a comprehensive industrial hygiene study of selected floors. Sixteen of 29 government-occupied floors were selected for evaluation. The building was divided according to the source of outdoor air supplied (roof or street level). Within each upper and lower half of the building, eight floors were selected for study: two floors of high occupant density with ninety or more employees and two floors of low occupant density with sixty or fewer employees, and four floors chosen at random. Lists of all employees assigned to the selected floors were compiled. A fifty percent random sample of these employees was then selected to participate in the questionnaire survey.

The industrial hygiene survey was conducted on the 8 floors selected by occupant density and outdoor air source. A primary objective of this survey was to compare the quality of air inside the building with that outside the building. This was accomplished by monitoring air contaminants in return air streams vs. contaminant levels detected in outdoor makeup air. For this reason most air samples were collected in the air handler mechanical rooms and at the face of the outdoor makeup air supply ducts. Air samples were collected for total airborne particulates, formaldehyde, and organic vapors. Carbon monoxide and carbon dioxide were monitored using length of stain detector tubes. Total hydrocarbons were measured using a H-Nu photoionization hydrocarbon detector. Settled dust samples and bulk samples of insulation were collected for microscopic identification of composition. Bag air samples were collected and qualitatively analyzed using a Wilks Model 1A infrared analyzer.

The average of a 6 point measurement of air flow at the outdoor makeup air duct opening was multiplied by the duct opening area to determine the outdoor air supply volume for each floor studied. This volume was later compared with ventilation guidelines recommended by the American Society of Heating, Refrigerating, and Air-Conditioning Engineers (ASHRAE), as published in ASHRAE Standard 62-1981, "Ventilation for Acceptable Indoor Air Quality."

Results

Questionnaire survey

Of the 493 questionnaires distributed, 327 were returned, providing a response rate of 66.3%. Although no serious illness was reported, the survey documented widespread complaints with 31.6% of the respondents reporting eye irritation,

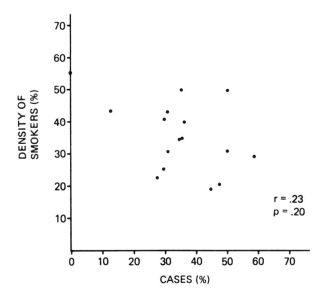

Figure 1 — Relationship of density of smokers to cases.

26.8% reporting sinus congestion, and 23.5% complaining of headache. Occupants who experienced one or more of these symptoms while at work at least twice a week were defined as cases. Of the 327 responding to the questionnaire, 113 were classified as cases, for an overall attack rate of 34.5%.

An evaluation of the questionnaire responses found that:

1. There were no differences between cases and non-cases with respect to age, education level, or race.
2. Females had a higher rate of reporting symptoms (79 cases out of 180 female respondents, 43%; 34 cases out of 147 male respondents, 23.1%).
3. By job category secretaries had the highest rate of reporting headache. There were no differences in attack rates for sinus congestion or eye irritation between secretaries and other employees. Secretaries who operated photocopier machines 10% or more of their time everyday did not report more symptoms than other secretaries.
4. Persons with a history of allergy or persons who wore contact lenses had higher attack rates for eye irritation and sinus congestion.
5. There was no correlation between the attack rates among respondents on the surveyed floors and the percentage of respondents from those floors who were smokers or who operated photocopiers (Figures 1 and 2.)
6. There was a slightly negative correlation when comparing attack rates by floor with the quality of outdoor air provided to each floor (Figure 3).

Ventilation measurements

The amount of outdoor air supplied per person per minute was generally adequate when compared with ASHRAE Standard 62-1981, which recommends a minimum of 20 cubic feet per minute per person for general offices where smoking is permitted (Table I). Indoor temperature and relative humidity readings were also within ASHRAE recommendations.

Indoor air contamination

In Table II air contaminants detected inside the building are compared with airborne contaminants found outside the building. As you would expect the number and concentration of detectable air contaminants inside the building is slightly higher than levels detected in outdoor air. However, those

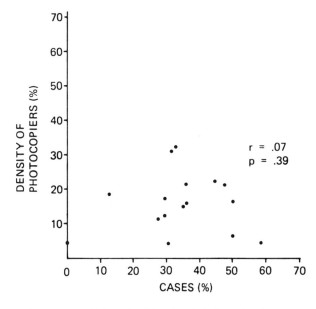

Figure 2 — Relationship of density of photocopiers to cases.

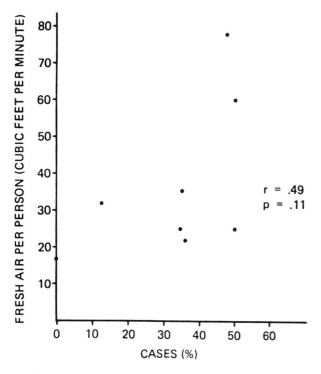

Figure 3 — Relationship of fresh air per person to cases.

indoor air contaminants were only found at very low concentrations well below levels which would be expected to cause adverse health effects for building occupants. Furthermore, none of the contaminants were above the more stringent 62-1981 ASHRAE guidelines or Georgia ambient air quality standards. For example, total airborne particulates inside the building ranged from 27-59 $\mu g/m^3$. This level was slightly higher than the concentration found in the outdoor air during the day of the survey. However, it is of little consequence considering that the Georgia Environmental Protection Division reported that total suspended particulate levels in the ambient air of downtown Atlanta averaged only 55 $\mu g/m^3$ during 1980. Although a greater variety of hydrocarbon contaminants were identified inside the building, concentrations detected were extremely low and in most samples the amounts found were just above the detection limits of our most sensitive sampling and analytical methods.

Infrared analyzer readings obtained from bag air samples of indoor and outdoor air did not identify any airborne contaminants. Some differ-

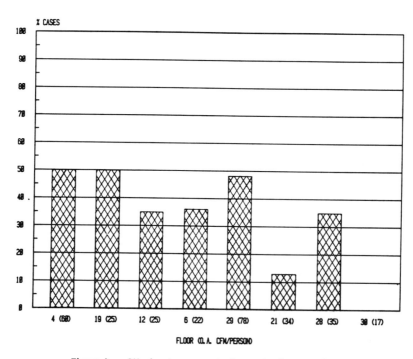

Figure 4 — Attack rates, percent of cases by floor number.

TABLE I
101 Marietta Tower Building, Atlanta, Georgia
TA 80-122, Ventilation Data
March 11-12, 1981

Floor No.	Sector	Fresh Air (cfm)	% Fresh Air	People/ Floor	Fresh Air/ Person (cfm)	Fresh Air/ sq ft (cfm)
4	E	1265	10.5			
4	W	580	4.8			
		1845		31	60	0.12
6	E	1297	10.8			
6	W	989	8.2			
		2286		103	22	0.15
21	W	728	6.1			
21	E	1085	9.0			
		1813		53	34	0.12
30	E	0	0			
30	W	1537	12.8			
		1537		92	17	0.10
12	W	1258	6.7			
12	E	1138	9.5			
		2396		94	25	0.16
19	E	969	8.1			
19	W	611	5.1			
		1580		63	25	0.11
28	E	1541	12.8			
28	W	1542	12.8			
		3038		88	35	0.21
29	E	1776	14.8			
29	W	1329	11.1			
		3105		40	78	0.21
Recommended minimum outdoor air (ASHRAE)						
non-smoking areas					5	0.035
smoking permitted		1050			20	0.14
		1050				
		2100				

E = East Sector
W = West Sector

% fresh air based on amount of fresh air provided (as measured by air velocity meter × air supply vent area) vs. designed cfm of recirculated air (12000 cfm per air handler).

Fresh air per sq. ft. of floor area = total fresh air ÷ 15000 sq ft/floor.

TABLE II
Air Contaminants

Found Outside	Found Inside
Total Hydrocarbons	Total Hydrocarbons
(HNU reading 2.0-3.8) C_9-C_{13} alkanes (roof and street) toluene (street)	(HNU reading 3.2-7.2) C_9-C_{13}; alkanes; toluene; benzene; perchloroethylene; 1,1,1-trichloroethylene; 120 M.W. aromatics; xylene
Airborne Particulates	Airborne Particulates (27-59 $\mu g/m^3$)
Roof air (none detected) Street level air (N.D-37 $\mu g/m^3$)	Found: cellulose fibers; soil particles (clay/quartz); errugular particles (ash)
	Not found: asbestos; fiberglass; mineral fibers
Carbon monoxide (0.5-1.0 ppm) Carbon dioxide (< 500 ppm) Formaldehyde (none detected)	Carbon monoxide (2.5-5.0 ppm) Carbon dioxide (< 500-900 ppm) Formaldehyde (N.D.-0.09 ppm)
NIOSH Exposure Limits/ACGIH TLVs	Alkanes 350 $\mu g/m^3$ Total Particulates (10,000 $\mu g/m^3$) Carbon monoxide (35 ppm) Carbon dioxide (50,000 ppm) Formaldehyde (0.5 ppm)
ASHRAE 62-1981 Guidelines	Carbon monoxide (35 ppm) Carbon dioxide (2500 ppm) Formaldehyde (0.1 ppm)
Georgia Ambiant Air Quality Standards 1980 Average for downtown Atlanta	Total Particulates (75 $\mu g/m^3$) Total Particulates (55 $\mu g/m^3$)

Other substances sampled for but not detected: ozone; sulfur dioxide; oxides of nitrogen

---

ences observed between indoor and outdoor air were attributed to fluctuations in water vapor and carbon dioxide content.

Discussion

As I previously mentioned, we found a slight negative correlation between attack rates and the quantity of outdoor air supplied. As displayed in Figure 4, the percentage of cases identified on floors where ventilation rates were measured were not affected by the level of outdoor air supplied. For example, floor 29 with an attack rate of 48% had the greatest supply of outdoor air (78 cfm/person). The floor with the lowest amount of outdoor air (floor 30 with 17 cfm/person) had the lowest attack rate (0%). In fact, the 30th floor had one of two outdoor air dampers fully closed at the time of the NIOSH survey. There was also no association found between attack rates and floor occupant density or source of the outdoor makeup air (roof vs street level air intake).

A further review of the environmental data for other possible relationships finds the amount of outdoor air provided to a floor may have some effect on the level of indoor air contaminants (Figure 5). With respect to airborne particulate levels you will note that the two floors with the highest airborne dust levels (floor 12 and 30) had relatively low outdoor air volumes (25 and 17 cfm/person respectively). The amount of outdoor air may also have affected the level of total indoor hydrocarbons detected on a particular floor (Figure 6). For example, one of the floors with a higher level of hydrocarbons, floor 30, also had the lowest outdoor air volume. However, I must point out that with only 8 floors surveyed during this investigation there is not sufficient data upon which to formulate any definite relationships. For instance, the relatively high level of hydrocarbons found on floor 21

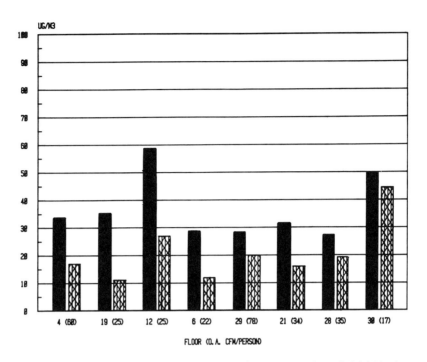

Figure 5 — Total particulate levels, indoor (solid bar) vs. outdoor air (plaid bar).

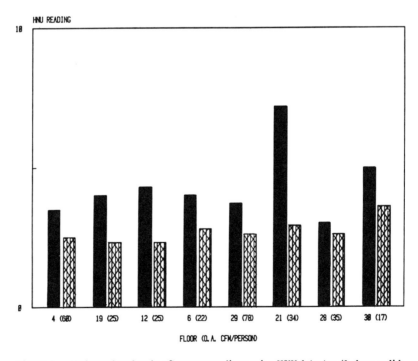

Figure 6 — Hydrocarbon levels, afternoon readings using HNU detector. (Indoor, solid bar; outdoor, plaid bar.)

can not be explained by a lack of adequate outdoor air which was found to be approximately 34 cfm/person at the time of the NIOSH survey.

Conclusions

If there is anything to be gained from this presentation my hope is that you will realize that use of conventional industrial hygiene approaches for conducting indoor air quality investigations will in most cases be nonproductive. The concentration of any specific indoor air contaminants detected will in almost all situations be far below levels which would explain the symptoms reported. In other words, air sampling should be conducted last, not first. If there is clear evidence that a high number of health complaints are occurring in an office or building, the first effort should be directed toward understanding and evaluating the operation of the building's HVAC system. If this effort determines that sufficient quantities of outdoor air are being provided, only then should air sampling be considered. The locations and methods for sampling should be carefully considered and if possible based on the results of administering a questionnaire to the building occupants. This questionnaire could be self-administered as long as occupants are asked not to discuss the contents of the questionnaire with fellow employees. If on the other hand, the building's HVAC is not providing adequate outdoor air, the system should be adjusted or modified to at least meet the ASHRAE 62-1981 guidelines. If possible, a follow-up questionnaire survey should be conducted one or two weeks after correcting the HVAC system. The results from this survey could be used to determine if changes made to the HVAC were effective in reducing occupant complaints.